Collins Revision

NEW GCSE

MATHS

Foundation

Revision Guide

For GCSE Maths from 2010

Edexcel + AQA + OCR

D1465242

Keith Gordon

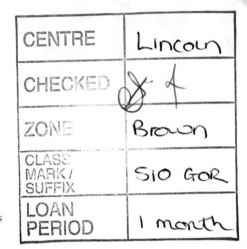

Published by Collins
An imprint of HarperCollins*Publishers*
77–85 Fulham Palace Road
Hammersmith
London W6 8JB

Browse the complete Collins catalogue at
www.collinseducation.com

10 9 8 7 6 5 4 3 2 1

ISBN-13 978-0-00-734098-9

British Library Cataloguing in Publication Data
A catalogue record for this publication is available from the British Library.

Written by Keith Gordon
Edited by Christine Vaughan
Project Managed by Philippa Boxer
Design by Graham Brasnett
Illustrations by Kathy Baxendale and Gray Publishing
Index compiled by Marie Lorimer
Printed by Mohn media Mohndruck GmbH in Germany

Collins Revision

NEW GCSE
MATHS
Foundation

Revision Guide

Keith Gordon

Contents

If you were in Year 10 or below in September 2010, you will be taking the new GCSE examination. The first one is in June 2012 but in fact, because all the examination boards are offering modular courses, you could be taking a modular examination as early as November 2010!

Each examination board has a slightly different way of organising their modules and different content is assessed in each of them, so you need to ask your teachers what examination you will be doing. There are three examination boards in England: AQA, EDEXCEL and OCR.

Once you know the examination you will be taking you can identify the sections to revise for any module by looking at top of each page in this revision guide: the modules each topic is assessed in by each board are listed. You will see that some of the topics appear in more than one module. This is because some topics in mathematics – such as basic number and basic algebra – are needed for all aspects of mathematics. Your teachers will be able to give you more guidance on the specific topics to be revised for each modular examination.

Remember, this is a revision guide and it is not intended to teach you any mathematics. Use it to remind you of the mathematics you have already learned. It will also give you some useful tips about the ways that certain topics are assessed in the examination.

To get the most from your revision, follow these steps:

- Read each page carefully.
- Work through the examples, checking that you understand and agree with each step.
- Read the 'Remember' bubbles – these give you advice on how to prepare for the examination.
- Work through the questions at the bottom of each page.
- Check your answers.
- If you are confident that you have mastered the topic, try the questions in the workbook.
- Check the answers to the workbook questions – have you have given the full answer that the examiners are looking for?

Each page of this revision guide is structured the same way.

Title: This tells you what the page is about.

Subtitle: These break the main topic up into different parts.

Grade: This tells you the top grade the topic could be assessed at.

Worked examples: These show you typical calculations using the facts and formulae.

Module guide: This tells you which module this topic will appear in for each board.

Remember bubble: This gives some useful tips about preparing for the exam.

Text: This tells you all the relevant facts about the topic that you need to know for the examination.

Questions: Typical examination questions with the grade they are targeted at.

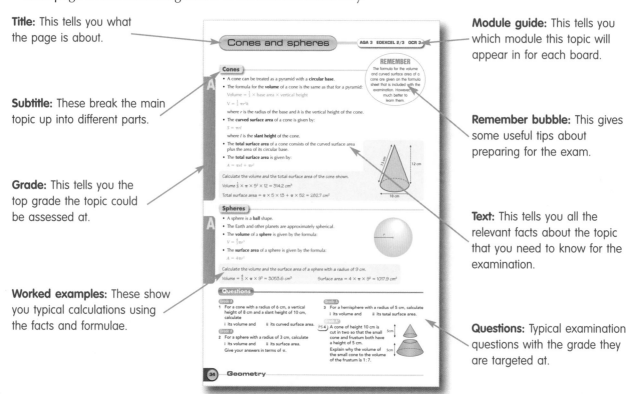

The revision guide has corresponding write-in workbook pages and the answers are in a detachable section at the back. The workbook also gives the grades of questions. When you check your answers, you will see that the mark scheme breaks the marks down so that you can see if you showed the necessary working.

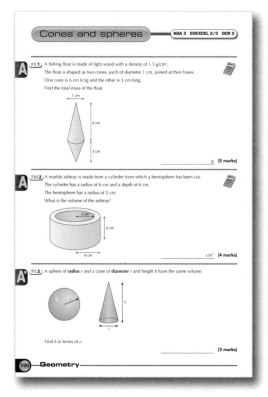

From September 2010 there will be a new mathematics GCSE. The first certificates will be awarded in June 2012.

For the English Awarding Bodies (AB) there are six specifications available.

AQA Mathematics Linear

This is a linear course – you will take two papers, one non-calculator and one calculator, in the same examination session. Each paper assesses all the topics in the specification, which means that you will get questions on Number and Algebra, Geometry and Measures, and Statistics and Probability in both papers.

AQA Mathematics Modular

This is a three-unit modular course.

Unit 1 is Statistics and Number

Unit 2 is Number and Algebra

Unit 3 is Geometry and Algebra

EDEXCEL Specification A

This is a linear course – you will take two papers, one non-calculator and one calculator, in the same examination session. Each paper assesses all the topics in the specification, which means that you will get questions on Number, Algebra, Geometry, Measures, Probability and Statistics in both papers.

EDEXCEL Specification B

This is a three-unit modular course.

Unit 1 is Statistics and Probability

Unit 2 is Number, Algebra, Geometry 1

Unit 3 is Number, Algebra, Geometry 2

OCR Specification A

This is a linear course – you will take two papers, one non-calculator and one calculator, in the same examination session. Each paper assesses all the topics in the specification, which means that you will get questions on Statistics, Number, Geometry and Measures, and Algebra in both papers.

OCR Specification B

This is a three-unit modular course. All three units cover Statistics, Number, Algebra, and Geometry and Measures.

Find out which course you are taking

Your teacher will be able to tell you which topics are assessed in each module, or you can find out from the examination boards' websites.

If you are taking a linear course, you will need to revise all the topics in this book to prepare for the final examination. If you are taking a modular course you will need to revise only the topics in each module before each examination. For example, you might do Module 1 in November of Year 10, Module 2 in June of Year 10 and Module 3 in June of Year 11. Look at the top of the pages in this book to see which modules a topic is assessed in.

What's new?

The new GCSE has some new types of question that haven't appeared in the GCSE examination before.

Functional mathematics

This is all about 'real-life' mathematics. They are not flagged on the examination paper, but in this revision guide and workbook functional questions are labelled with (FM). One way of deciding if a question is functional is to decide if anyone would want to know the information in a real-life context.

For example:
'Work out 30% of £150' is not functional but 'Work out the new price of a suit normally costing £150 after a 30% reduction in a sale' is a functional question.

In the Foundation Paper, 40% of the questions will have a functional mathematics element. In the Higher paper, 20% will have a functional element. Many functional mathematics questions will be very familiar to you (you've probably come across questions like the one above about the suit reduced in a sale in past papers) but some may be new. Unless you have done a Functional Mathematics Certificate, you may not have come across this sort of question before, for example:

Example (Grade D)

Julie belongs to a local gym.

This table shows the calories used on four types of exercise machines at the gym.

Exercise machine	Number of calories per minute		
	Intensity		
	Low	Med	High
Rowing machine	8	11	14
Treadmill	6	9	12
Static bicycle	5	8	11
Cross-trainer	10	14	18

Julie goes to the gym for an hour.

Julie likes to do about 30 minutes of low-intensity exercise and no more than 10 minutes of high-intensity exercise.

Devise a training programme that uses each machine and allows Julie to use a total of at least 500 calories. **[5 marks]**

Answer

This is an **open-ended** question – it does not have a definite answer.

It is important to make sure that you communicate your answer to the examiner and do enough mathematics to get all the marks available.

This would be a good answer:

15 minutes on the rowing machine at low intensity will be $15 \times 8 = 120$ calories used.

10 minutes on the cross-trainer at high intensity will be $10 \times 18 = 180$ calories used.

20 minutes on the static bicycle at medium intensity will be $20 \times 8 = 160$ calories used.

15 minutes on the treadmill at low intensity will be $15 \times 6 = 90$ calories used.

Total calories $= 120 + 180 + 160 + 90 = 550$ calories.

There are many other possible answers: the examiners will not be looking for a single 'right' answer – instead they will be checking that all the conditions are met and that the calculations are correct.

For example:

Are all machines used? ✓   1 hour in total? ✓

10 minutes at high intensity? ✓   At least 500 calories? ✓

30 minutes at low intensity? ✓   All calculations correct? ✓

Assessment Objective 1 (AO1)

About half of the questions in the examination will be straightforward questions that test if you can do mathematics. These are known as Assessment Objective 1 (AO1) questions. The examples below are AO1 questions, with advice on how to answer them. You will have come across this sort of question often.

Example (Grade D)

In a school there are 600 students and 30 teachers.

12% of the students are vegetarian.

10% of the teachers are vegetarian.

How many vegetarians are there in the school altogether?   **[3 marks]**

Answer

$0.12 \times 600 = 72$ students are vegetarian | Use the percentage multipliers 0.12 and 0.1 to work out 12% and 10%. |

$0.1 \times 30 = 3$ teachers are vegetarian

$3 + 72 = 75$ total vegetarians | Show the total clearly. |

Make sure that all calculations are shown. It is always a good idea to use percentage multipliers in this sort of question as it makes calculations easier.

Example (Grade D)

Solve the equation $6x + 5 = 9 - 2x$ **[3 marks]**

Answer

$6x + 5 = 9 - 2x$

$6x + 2x = 9 - 5$

$8x = 4$

$x = 0.5$

Assessment Objective 2 (AO2)

About 30% of the examination questions are designed to test whether you understand the topics and can apply mathematics in slightly more involved situations. These are known as Assessment Objective 2 (AO2) questions. They are similar to the old 'Using and Applying Mathematics' questions, except that they are designed to test your **understanding**, not just whether you can do the mathematics. They are not flagged in the examination, but in this revision guide and workbook AO2 questions are labelled with (AU). The examples opposite are AO2 questions, with advice on how to answer them.

Example (Grade E)

Here are the instructions for cooking a joint of beef.

Cook for 20 minutes at 200 °C.

Reduce the temperature to 160 °C and cook for 10 minutes per pound.

Kevin has a 6 pound joint of beef that he wants to take out of the oven at 1 pm.

What time should he start to cook it? **[3 marks]**

Answer

Time to cook $= 20 + 6 \times 10$

$= 80$ minutes

> Show the calculation for working out the cooking time. This will get a method mark. The time of 80 minutes will get an accuracy mark.

Time to put in oven is 80 minutes before 1pm

$= 11.40$ am

> Work out what time is 80 minutes before 1 pm.

Example (Grade D)

An isosceles triangle has one angle of 70°.

Work out what the other angles could be. **[3 marks]**

Answer

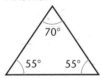

> Drawing a diagram will get full marks as will writing out the values.

One possible triangle has angles of 70°, 55° and 55°.

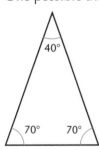

> The question doesn't tell you which angle is 70°. It could be the equal angle or it could be the non-equal angle, so show both possibilities.

One possible triangle has angles of 40°, 70° and 70°.

Assessment Objective 3 (AO3)

About 20% of the examination questions are designed to test whether you can solve problems. These are known as Assessment Objective 3 (AO3) questions. They are not flagged in the examination, but in this revision guide and workbook AO3 questions are labelled with (PS). The following examples are AO3 questions, with advice on how to answer them.

Example (Grade C)

At the school summer fair, Ray sets up a stall to raise money.

Players pay 50p to throw two dice.

If they throw two 6s they get £2 back. If they throw one 6 they get £1 back.

If 100 people play the game, how much money would Ray expect to raise?

You may use the grid on the next page to help you. **[4 marks]**

+	Score on first dice					
	1	2	3	4	5	6
1						
2						
3						
4						
5						
6						

(Score on second dice)

Answer

$P(\text{two 6s}) = \dfrac{1}{36}$

> Use the table to work out the probabilities.

$P(\text{one 6}) = \dfrac{10}{36}$

Number of two 6s in 100 throws $= \dfrac{1}{36} \times 100 \approx 3$ which is £6 paid out

> Use the probabilities to work out the expected values.

Number of one 6 in 100 throws $= \dfrac{10}{36} \times 100 \approx 28$ which is £28 paid out

Income = £50

> Work out the expected profit.

Profit = £50 − £28 − £6 = £16

Example (Grade D)

Tammy makes three-legged stools and four-legged tables.

The stools sell for £12 and the tables sell for £30.

One day she sells enough stools and tables to make £180 and uses 31 legs.

How many stools and tables did she sell?   **[3 marks]**

Answer

1 table costs £30 which leaves £150 which does not divide by 12.

> Assume one table is sold, which does not work.

2 tables cost £60 which leaves £120 which would be 10 stools.

> Assume two tables are sold, which leaves 10 stools.

This would be $2 \times 4 + 10 \times 3 = 38$ legs.

> Work out how many legs this is.

3 tables cost £90 which leaves £90 which does not divide by 12.

> Keep on increasing the number of tables.

4 tables cost £120 which leaves £60 which would be 5 stools.

This would be $4 \times 4 + 3 \times 5 = 31$ legs.

> Eventually the number of table and stools will give the correct number of legs.

Quality of written communication

Some questions in the examination are designed to assess how well you set out and explain your answers, or your Quality of Written Communication (QWC). All examination boards will indicate in some way which questions are allocated marks for QWC.

This is another new requirement for mathematics, although it has been assessed in other GCSE subjects for some time.

To earn QWC marks, you will need to:

- write legibly, with accurate use of spelling, grammar and punctuation in order to make your meaning clear
- select and use a form and style of writing appropriate to purpose and to complex subject matter
- organise relevant information clearly and coherently, using specialist vocabulary when appropriate.

To help you do this, think about whether what you have written will make sense to someone else. Make sure your answer is clear and logical. This is a good idea anyway, because if the examiner can understand your method, you might get some marks even if your final answer is wrong.

You should also make sure that you use correct mathematical notations, such as £3.60 not £3.6, or $5x$ not $x5$, and use specialist vocabulary: for example, say 'because it is an alternate angle' rather than 'it is a Z angle'.

Here is an example of a question that has a QWC mark allocated to it.

Example (Grade E)

This bar chart shows the number of boys and girls who have a salad for school lunch in one week.

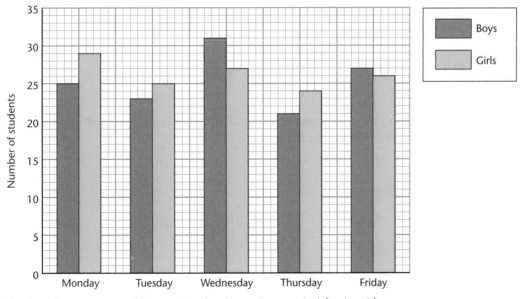

The headteacher says 'More girls than boys have salad for lunch'.

Is the headteacher correct?

Justify your answer.

[3 marks]

Mark scheme

Answer	Mark
Working out the total for boys or girls	M1
Correct totals for boys (127) **and** girls (131)	A1
Valid conclusion using data e.g. Head is correct as $131 > 127$	QWC1

Answer

Here are three answers.

127

131

Yes $131 > 127$

> This would just about get three marks as it has all the necessary information, but there is not much communication in this answer.

Total for boys = 136

Total for girls = 131

The head is wrong as more boys have salad as $136 > 131$

> This answer would get two marks out of three (M1, A0, QWC1). Even though the total for the boys is wrong, the total for the girls is correct, which earns the method mark (M1). The conclusion is correct for the values calculated, earning the communication mark (QWC1). However, the total for boys is wrong, so the accuracy mark is zero (A0).

Total for boys = 136

Total for girls = 121

The head is wrong as more boys have salad as $136 > 121$

> This answer would get 1 mark out of three (M0, A0, QWC1) as both totals are wrong, but the conclusion is correct for the values calculated.

This shows that it is important to use words to communicate what you are doing so that you get the mark for communication even if you make mathematical mistakes.

Statistical representation

Statistics

- Statistics is concerned with the collection and organisation of **data**.

- It is usual to use a **data collection sheet** or **tally chart** for collecting data.

- Data is recorded by means of **tally marks**, which are added to give a **frequency**.

- Data can be collected in a **frequency table**.

This tally chart shows the results when Alf threw two coins together several times.

Outcome	Tally	Frequency
2 tails	⁄⁄⁄⁄ ⁄⁄⁄⁄ ⁄⁄⁄⁄	14
1 tail, 1 head	⁄⁄⁄⁄ ⁄⁄⁄⁄ ⁄⁄⁄⁄ ⁄⁄⁄⁄ ⁄⁄⁄⁄ ⁄⁄⁄	
2 heads	⁄⁄⁄⁄ ⁄⁄⁄⁄ ⁄⁄⁄⁄ ⁄⁄⁄	

Collecting data

- Data can be collected by three methods.

  - **Taking a sample**, for example, to find which 'soaps' students watch, ask a random selection of 50 students.

  - **Observation**, for example, to find out how many vehicles use a road in a day, count the vehicles that pass during an hour.

  - **Experiment**, for example, to see if a home-made spinner is biased, throw it about 100 times and record the outcomes.

> **REMEMBER**
> When collecting or tallying, always double-check as it is easy to miss a piece of data.

Pictograms

- A pictogram is a way of showing data in a diagrammatic form that uses **symbols** or **pictures** to show **frequencies**.

- Every pictogram has a **title** and a **key** that shows how many items are represented by each symbol.

> **REMEMBER**
> Always choose a symbol that is easy to draw and can be divided into equal parts.

Average daily number of hours of sunshine

Month	Hours of sunshine	Total
June	☼☼ ☽	11
July	☼☼ ☼☽	
August		13

Key: ☼ represents 4 hours

Questions

Grade G

1 a Copy the tally chart, above, and complete the frequency column.

  b Which outcome is most likely?

  c How many times altogether did Alf throw the two coins?

Grade G

FM 2 The pictogram, above, shows the average daily hours of sunshine for a town in Greece during June, July and August.

  a What was the daily average for July?

  b Complete the pictogram for August.

Bar charts

- A **bar chart** is made up of bars or blocks of the same width, drawn horizontally or vertically on an axis.

- The **heights** or **lengths** of the bars always represent **frequencies**.

- The bars are separated by small gaps to make the chart easier to read.

- Both **axes** should be **labelled**.

- The **bar chart** should be **labelled** with a **title**.

- A **dual bar chart** can be used to compare two sets of data.

This graph shows sales of two newpapers over a week.

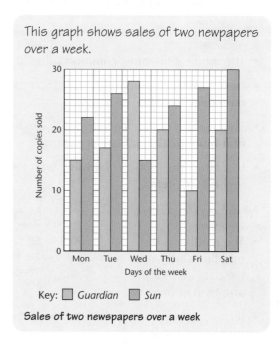

Sales of two newspapers over a week

Line graphs

- **Line graphs** are used to show how data changes over a period of time.

- Line graphs can be used to show **trends**, for example, how the average daily temperature changes over the year.

- **Data points** on line graphs can be joined by **lines**.
  - When the lines join points that show **continuous data**, for example, joining points showing the height of a plant each day over a week, they are drawn as **solid lines**. This is because the lines can be used to estimate intermediate values.
  - When the lines join points that *do not* show continuous data, for example, the daily takings of a corner shop, they are drawn as **dotted lines**. This is because the lines cannot be used to estimate values, they just show the **trend**.

This graph compares temperatures.

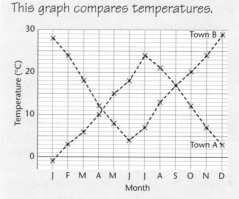

Temperatures in a town in Australia and a town in Britain

REMEMBER
The largest increase or decrease on a line graph is at the steepest part of the graph.

Questions

Grade F

FM 1 Refer to the dual bar chart, above.

a On which two days were the same number of copies of the *Guardian* sold?

b How many copies of the *Sun* were sold altogether over the week?

c On which day was there the greatest difference between the numbers of copies sold of the *Guardian* and the *Sun*?

d On one day the *Guardian* had a 'half-price' offer. Which day was this? Explain your answer.

Grade D

FM 2 Refer to the line graph, above.

a Which town is hotter, on average? Give a reason for your answer.

b Which town is in Australia? Give a reason for your answer.

c In which month was the average temperature the same in both towns?

d Is it true that the average temperature was the same in both towns on a day in early April?

Statistics

Averages

AQA 1   EDEXCEL 1   OCR 1

- An **average** is a single value that gives a **representative value** for a set of data. There are three types of average: **mode**, **median** and **mean**.

Mode

D

- The **mode** is the **most common** value in a set of data.
- Not all sets of data have a mode. Some data sets have **no mode**.
- The mode is the only average that can be used for **non-numerical** (**qualitative**) data.
- In a table, the mode is the **value** with the **highest frequency**.

> These are the numbers of eggs in 20 sparrows' nests.
>
> 2, 4, 1, 3, 5, 1, 3, 1, 4, 5,
> 1, 2, 3, 3, 1, 5, 2, 1, 3, 2
>
> The **modal number** of eggs per nest is 1.

Median

D

> **REMEMBER**
> Don't forget to put the data in order, to find the median.

- The **median** is the **middle value** when the data items are arranged in order.
- If there are n pieces of data the median is the $\frac{n+1}{2}$ th number in the list.
- If the number of items of data is **odd**, the middle value will be one of the data items.
- If the number of items of data is **even**, the middle value will be halfway between two of the data items.
- To find the median in a **frequency table**, add up the frequencies in the table until the total passes half the number of data items in the whole set.

> This frequency table has been used to record results of a survey.
>
> | Value | 3 | 4 | 5 | 6 | 7 | 8 |
> |---|---|---|---|---|---|---|
> | Frequency | 7 | 8 | 9 | 12 | 3 | 1 |
> | Total frequency | 7 | 15 | 24 | | | |
>
> There are 40 values so the median is the $\frac{40+1}{2} = 20\frac{1}{2}$th value.
>
> Add up the frequencies until you find where this value will be.
>
> The median must be in the third column, so the median is 5.

Mean

C

- The mean is defined as $\text{mean} = \dfrac{\text{sum of all values}}{\text{number of values}}$
- When people talk about 'the average' they are usually referring to the mean.
- The advantage of using the mean is that it **takes all values into account**.

> These are the numbers of eggs in 20 sparrows' nests.
>
> 2, 4, 1, 3, 5, 1, 3, 1, 4, 5,
> 1, 2, 3, 3, 1, 5, 2, 1, 3, 2
>
> The **mean number** of eggs per nest is $52 \div 20 = 2.6$.

Questions

Grade D

1 a Find the mode for each set of data.

   i 4, 5, 8, 4, 3, 5, 6, 4, 5, 7, 9, 5, 3, 8

   ii Y, B, Y, R, G, Y, R, G, Y, Y, R, B, B, Y, R

  AU b Why is the modal number of eggs in sparrows' nests, above, not a good average?

Grade D

2 Find the median for each set of data.

  a 8, 7, 3, 2, 10, 8, 6, 5, 9

  b 5, 13, 8, 7, 11, 6, 12, 9, 15, 4, 3, 10

Grade C

3 Use a calculator to find the mean of each data set.

  a 44, 66, 99, 44, 34, 66, 83, 70, 45, 76, 77

  b 34.5, 44.8, 29.3, 27.2, 34.1, 39.0, 30.4, 40.7

Grade C

PS 4 Find five numbers with a mean of 7, a mode of 8 and a range of 9.

Averages and range

Range

- The **range** of a set of data is the **highest value minus the lowest value**.

  Range = highest value − lowest value

- The range is used to measure the **spread of data**.

- The range is used to comment on the **consistency of the data**.
  A smaller range indicates more consistent data.

> **REMEMBER**
> Remember! The range is *not* an average.

Which average to use

- The average must be truly **representative** of the data, so the average used must be **appropriate** for the set of data.

- The table below shows the advantages and disadvantages of each average.

	Mode	Median	Mean
Advantages	• Easy to find	• Easy to find for ungrouped data	• Uses all values
	• Not affected by extreme values	• Not affected by extreme values	• Original total can be calculated
Disadvantages	• Does not use all values	• Does not use all values	• Most difficult to calculate
	• May be at one extreme	• Raw data has to be put in order	• Extreme values can distort it
	• May not exist	• Hard to find from table	
Use for	• Non-numerical data	• Data with extreme values	• Data with values that are spread in a balanced way
	• Data sets in which a large number of values are the same		• Data in which all values are relevant, for example, cricket score averages

Questions

Grade F

1 Find the range of each of these sets of data.

   a 4, 9, 8, 5, 6, 10, 11, 7, 8, 5
   b −4, 8, 5, −1, 0, 0, 2, 4, 3, −2, −3, 4

Grade F

FM 2 For a darts match, the captain has to choose between Don and Dan to play a round. During the practice, the captain records the scores they both make with three darts on five throws. These are:

   Dan 120, 34, 61, 145, 20
   Don 68, 89, 80, 72, 71

   a Work out the mean for each player.
   b Work out the range for each player.
   c Who should the captain pick?
      Explain why.

Grade C

AU 3 After a maths test, the class were told the mean, mode and median marks.

   Three students made these statements.
   Asaf: 'I was in the top half of the class.'
   Brian: 'I can't really tell how well I have done.'
   Clarrie: 'I have done really well compared with the rest of the class.'

   Which averages did they use to make these statements?

Grade C

AU 4 These are the weekly wages in a small printing company:
   1 apprentice £100
   5 printers   £300, £300, £300, £300, £300,
   1 manager   £400
   1 director   £800

   Which average would best represent the weekly wage in the company?

Statistics

Arranging data

D

Frequency tables

- When you have to represent a lot of data, use a **frequency table**.

This table shows how many times students in a form were late in a week.

Number of times late	0	1	2	3	4	5
Frequency	11	8	3	3	3	2

> **REMEMBER**
> The mode is the value that has the highest frequency; it is not the actual frequency.

- The **mode** is the data value with the **highest frequency**.

- Find the **median** by adding up the frequencies of the data items, in order, until the halfway point of all the data in the set is passed.

- Find the **mean** by multiplying the value of each data item by its frequency, adding the totals, then dividing by the total of all the frequencies.

> To find the mean number of times students were late, work out the total number of times students were late: $0 \times 11 + 1 \times 8 + 2 \times 3 + 3 \times 3 + 4 \times 3 + 5 \times 2$ and work out the total frequency: $11 + 8 + 3 + 3 + 3 + 2$, then divide the first total by the second.

C

Grouped data

- A wide range of data, with lots of values, may have too many entries for a frequency table, so use a **grouped frequency table**.

- Grouped data can be shown by a **frequency polygon**.

- For a frequency polygon, plot the **midpoint** of each group against the **frequency**.

- In a grouped frequency table, data is **recorded in groups** such as $10 < x \leq 20$.

- $10 < x \leq 20$ means values between 10 and 20, not including 10 but including 20.

- The **modal class** is the group with the greatest frequency. It is not possible to identify the actual mode.

- The **median** cannot be found from a grouped table.

- Calculate an **estimate of the mean** by adding the midpoints multiplied by the frequencies and dividing the result by the total frequency.

This table shows the marks in a mathematics examination for 50 students. The frequency polygon shows the data.

Marks, x	Frequency, f
$0 < x \leq 10$	4
$10 < x \leq 20$	9
$20 < x \leq 30$	17
$30 < x \leq 40$	13
$40 < x \leq 50$	7

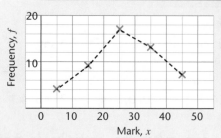

Questions

Grade D

FM 1 Study the table showing how often students were late in a week.

   a How many students were there in the form?

   b What is the modal number of times they were late?

   c **i** What is the total number of 'lates'?

     ii What is the mean number of 'lates' per day?

Grade C

FM 2 The marks for 50 students in a mathematics examination are shown in the table, above.

   a What is the modal class?

   b **i** What is the total of the 'midpoints times frequencies'?

     ii What is the estimated mean mark for the form?

Statistics

Stem-and-leaf diagrams

- When data is first recorded, it is called **raw data** and is **unordered**.

  The ages of the first 15 people to use a shop in the morning were:

  18, 26, 32, 29, 31, 57, 42, 16, 23, 42, 30, 19, 42, 35, 38

- Unordered data can be put into order to make it easier to read and understand. This is called **ordered** data.

- A **stem-and-leaf diagram** is a way of showing ordered data.

  This stem-and-leaf diagram shows the marks scored by students in a test.

  ```
  1 |  3  4  5  6
  2 |  0  1  1  3  4  5  8
  3 |  1  2  5  5  5  9
  4 |  2  2  9
  ```
  Key 1 | 3 represents 13 marks

  > **REMEMBER**
  > Always check the key of a stem-and-leaf diagram, in case the 'stem' is not tens.

- The **stem** is the number on the left of the vertical line. In this case, it represents the tens.

- The **leaves** are the numbers on the right of the vertical line. In this case, they represent the units.

- A stem-and-leaf diagram must have a **key**, which shows what the stem and the leaves stand for. The stem is *not always* tens.

- In the diagram above the first six entries are 13, 14, 15, 16, 20 and 21.

- Find the **mode** by finding the most common entry.

- Find the **median** by counting from the start to the middle value.

- Find the **range** by subtracting the lowest value from the highest value.

- Find the **mean** by adding all the values and dividing by the total number of values in the table.

Questions

Grade C

1 The ages of the first 15 people to use a shop in the morning are shown above.

  a Rearrange the raw data into an ordered list and draw a stem-and-leaf diagram.

  b What is the modal age?

  c Why is the mode not a good average to use for this data?

  d What is the median age?

  e What is the range of the ages?

  f What is the mean age?

Grade C

FM 2 The marks for some students in a test are shown in the stem-and-leaf diagram, above.

  a How many students took the test?

  b i What is the lowest mark in the test?

   ii What is the highest mark in the test?

   iii What is the range of the marks?

  c What is the modal mark?

  d What is the median mark?

  e i What is the total of all the marks?

   ii What is the mean mark?

Statistics

Probability

The probability scale

- People use everyday words such as '**chance**', '**likelihood**' and '**risk**' to assess whether something will happen.

- **Probability** is the mathematical method of assessing chance, likelihood or risk.

- Probability gives a value to the **outcome** of an **event**.

- The **probability scale** runs from **0 to 1**.

- Terms such as '**very likely**' and '**evens**' are used to associate probabilities with approximate positions on the scale.

> 'What is the chance of rain today?' or 'Shall we risk not taking an umbrella?'

> The probability of throwing a six with a dice is one-sixth.

> **REMEMBER**
> Probabilities may be given as fractions or percentages but they can never be less than zero or greater than 1.

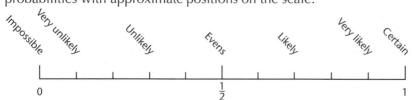

- The probability of an **impossible** event is 0.

- The probability of a **certain** event is 1.

> The probability that pigs will fly is 0.

> The probability that the sun will rise tomorrow is 1.

Calculating probabilities

- Whenever anyone does something, such as take a card from a pack of cards or buy a raffle ticket, there are a number of possible outcomes.

  There are 52 possible outcomes of picking a card from a pack of cards.

- There may be more than one way of any particular outcome of an event occurring.

  There are four possible ways of picking a king from a pack of cards.

- The probability of an outcome of an event is defined as:

  $$P(\text{outcome}) = \frac{\text{number of ways the outcome can occur}}{\text{total number of possible outcomes}}$$

- At random means 'without looking' or 'not knowing the outcome in advance'.

Questions

Grade G

1 a State whether each event is impossible, very unlikely, unlikely, evens, likely, very likely or certain.

  i Christmas Day being on 25 December.
  ii Scoring an even number when a regular dice is thrown.
  iii Someone in the class having a mobile phone.
  iv A dog talking.

b Draw a probability scale and mark an arrow for the approximate probability of each outcome.

  i The next person to walk through the door will be female.

  ii The person sitting next to you in mathematics is under 18 years of age.
  iii Someone in the class will have a haircut today.

Grade E

2 What is the probability of picking the following from a pack of cards?

  a A red card
  b A king
  c A red king

Grade C

3 What is the probability of each of these events?

  a Throwing a 6 with a dice.
  b Picking a picture card from a pack of cards.
  c Tossing a tail with a coin.

Using probability

Probability of 'not' an event

> **REMEMBER**
> When subtracting a fraction from 1, just find the difference between the numerator and the denominator for the new numerator, for example, $1 - \frac{3}{8} = \frac{(8-3)}{8} = \frac{5}{8}$.

- To find the probability that an event **does not happen**, use the rule:

  $P(\text{event}) + P(\text{not event}) = 1$ or $P(\text{not event}) = 1 - P(\text{event})$

  If the probability that a student picked at random from a class is a girl is $\frac{5}{11}$, then the probability that a student picked at random is not a girl (is a boy) is $1 - \frac{5}{11} = \frac{6}{11}$.

Addition rule for mutually exclusive outcomes

- Mutually exclusive outcomes or events cannot occur at the same time.

- For the probability of either of two mutually exclusive events, **add** the probabilities of the separate events: $P(A \text{ or } B) = P(A) + P(B)$

> **REMEMBER**
> This rule only works with mutually exclusive events.

  Picking a jack and a king from a pack of cards are mutually excusive events.

  The probability of picking a jack or a king is: $\frac{4}{52} + \frac{4}{52} = \frac{8}{52} = \frac{2}{13}$

  The probability of picking a king or a red card is not $\frac{4}{52} + \frac{26}{52}$ because there are two red kings.

Relative frequency

> **REMEMBER**
> If you are asked who has the best set of results, always say the person with the most trials.

- The probability of an event can be found by three methods.
  - **Theoretical probability** for **equally likely outcomes**, such as drawing from a pack of cards.
  - **Experimental probability**, calculating **relative frequency**, for example, from a large number of trial spins of a home-made spinner.
  - **Historical data**: looking up past records, for example, for an earthquake in Japan.

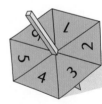

- To calculate the relative frequency of an event, **divide** the **number of times** the event occurred during the experiment by the **total number of trials**.

- The **more trials** that are done, the nearer the experimental probability will be to the true probability.

Questions

1 The probability that someone picked at random is left-handed is $\frac{3}{10}$. What is the probability that someone picked at random is not left-handed?

2 What is the probability that a card picked randomly from a pack is:

a an ace

b a king

c an ace or a king?

3 These are the results when three students tested the spinner shown above.

Student	Ali	Barry	Clarrie
Number of throws	20	60	240
Number of 4s	5	13	45

a For each student, calculate the relative frequency of a 4. Give your answers to 2 decimal places.

b Which student has the most reliable estimate of the actual probability of a 4? Explain why.

c If the spinner was fair, how many times would you expect it to land on 4 in 240 spins?

Statistics

Combined events

- When two events occur together they are known as **combined events**.
- The outcomes of the combined events can be shown as a list.

> If two coins are thrown together, the possible outcomes are: (head, head), (head, tail), (tail, head) and (tail, tail).

- Another method for showing the outcomes of a combined event is to use a **sample space diagram**.

If two dice are thrown, the possible outcomes can be shown like this.

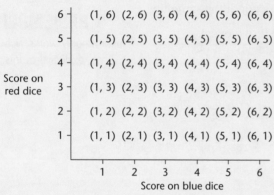

If two dice are thrown together and the scores are added, the possible outcomes can be shown like this.

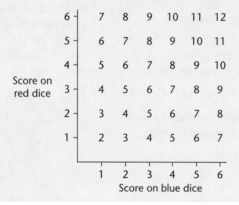

Questions

Grade B

1 a When two dice are thrown together, how many possible outcomes are there?

   b Refer to the top sample space diagram for throwing two dice, above.

   i What is the probability of throwing a double with two dice?

   ii What is the probability that the difference between the scores on the two dice is 4?

   c Refer to the bottom sample space diagram for throwing two dice, above.

   i What is the probability of throwing a score of 5 with two dice?

   ii What is the probability of throwing a score greater than 9 with two dice?

   iii What is the most likely score with two dice?

Expectation and two-way tables

Expectation

- When the probability of an event is known we can **predict** how many times the event is likely to happen in a given number of trials.

- This is the **expectation**. It is *not* what is going to happen.

> If a coin is tossed 1000 times we would expect 500 heads and 500 tails.
> It is very unlikely that we would actually get this result in real life.

- The **expected number** is calculated as: **Expected number = P(event) × total trials**

Two-way tables

- A two-way table is a table that links two variables.

This table shows the languages taken by the boys and girls in Form 9Q.

	French	Spanish
Boys	7	5
Girls	6	12

REMEMBER

Usually a column and a row for the totals are included in the table. If they are not, always add up each row and column anyway. You will almost always need these values to answer the questions.

- One of the variables is shown by the rows of the table.

- One of the variables is shown by the columns of the table.

This table shows the nationalities of people on a jet plane and the types of ticket they have.

	First class	Business class	Economy
American	6	8	51
British	3	5	73
French	0	4	34
German	1	3	12

Questions

Grade D

1 A bag contains 10 counters. Five are red, three are blue and two are white. A counter is taken from the bag at random. The colour is noted and the counter is replaced in the bag. This is repeated 100 times.

   a How many times would you expect a red counter to be taken out?

   b How many times would you expect a white counter to be taken out?

   c How many times would you expect a red or a white counter to be taken out?

Grade C

PS 2 Some more red and green counters are added to the bag of counters in question 1 so that the probability of taking a green counter at random is $\frac{1}{4}$. How many of each colour were added?

Grade D

3 a Refer to the two-way table for Form 9Q, above. Nobody takes two languages.

   i How many boys take Spanish?

   ii How many students are in the form altogether?

   iii How many students take French?

   iv What is the probability that a student picked at random from the form is a girl who takes Spanish?

   b Refer to the two-way table for the plane travellers, above.

   i How many travellers were there on the plane?

   ii What percentage of the travellers had first-class tickets?

   iii What percentage of the business-class passengers were American?

Statistics

Pie charts

Pie charts

- A **pie chart** is another method of representing data.
- Pie charts are used to show the **proportions** between different categories of the data.
- The **angle** of each sector (slice of pie) is **proportional** to the **frequency** of the category it represents.

This table shows the favourite colours of 20 pupils.

The angle is worked out by multiplying the frequency by 360 divided by the total frequency.

Colour	Frequency	Calculation	Angle
Red	4	4 × 360 ÷ 20	72°
Blue	7	7 × 360 ÷ 20	126°
Green	5	5 × 360 ÷ 20	90°
Yellow	4	4 × 360 ÷ 20	72°
		Total	360°

- Pie charts *do not* show individual frequencies, they only show proportions.
- Pie charts should always be **labelled**.

This pie chart shows the types of transport used by a group of people going on holiday.

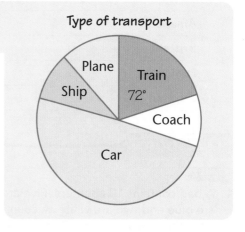

Type of transport

Questions

FM 1 Some people were asked what type of transport they mainly used on their holidays. Their replies are shown in the 'Type of transport' pie chart, above.

  a There were 24 people who replied 'Train'.

  How many people were in the survey altogether?

  b There were 11 people who replied 'Ship'.

  What is the angle of the sector representing 'Ship'?

2 This table shows the favourite colours of a class of 30 students.

Colour	Frequency
Red	12
Blue	5
White	3
Black	6
Green	4

Draw a fully labelled pie-chart to show the data.

Statistics

Scatter diagrams

D

Scatter diagrams

- A **scatter diagram** (also known as a **scattergraph** or **scattergram**) is a diagram for comparing two variables.

- The variables are plotted as **coordinates**, usually from a table.

Here are 10 students' marks for two tests.

Tables	3	7	8	4	6	3	9	10	8	6
Spelling	4	6	7	5	5	3	10	10	9	7

This is the scatter diagram for these marks.

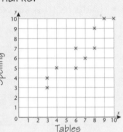

Correlation

- The scatter diagram will show a **relationship** between the variables if there is one.

- The relationship is described as **correlation** and can be written as a **'real-life' statement**.

For the first diagram: 'The taller people are, the bigger their arm span'.

Positive correlation

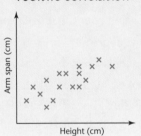

Height (cm)

Negative correlation

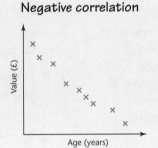

Age (years)

No correlation

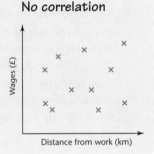

Distance from work (km)

C

Line of best fit

- A **line of best fit** can be drawn through the data.

- The line of best fit can be used to **predict** the value of one variable when the other is known.

> **REMEMBER**
> Draw the line of best fit between the points with about the same number of points on either side of it.

The line of best fit passes through the 'middle' of the data.

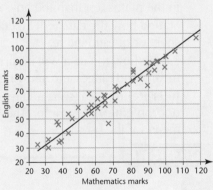

C

Questions

Grade D

FM 1 a Refer to the scatter diagram of students' scores for tables and spelling tests, above. What type of correlation does the scatter diagram show?

b Describe the relationship in words.

Grade C

FM 2 a Describe, in words, the relationship between the value of a car and the age of the car, shown in the second of the three scatter diagrams, above.

b Describe, in words, the relationship between the wages and the distance travelled to work, shown in the third of the three scatter diagrams, above.

Grade C

FM 3 Refer to the line of best fit on the scatter diagram showing the English and mathematics marks. Estimate the score in the English examination for someone who scored 75 in the mathematics examination.

Statistics

Surveys

- **A survey** is an organised way of finding people's opinions or testing an hypothesis.
- Data from a survey is usually collected on a **data collection sheet**.

This data collection sheet shows the favourite types of chocolate for 64 students.

Type of chocolate	Tally	Frequency
Milk	ﬀ ﬀ ﬀ ﬀ ﬀ ///	28
Plain	ﬀ ﬀ ﬀ //	17
Fruit and nut	ﬀ ﬀ /	11
White	ﬀ ///	8

- **Questionnaires** are used to collect a lot of data.
- **Questions** on questionnaires should follow some rules.
  - Never ask a **leading question** or a **personal question**.
  - Keep questions **simple**, with a **few responses** that **cover all possibilities** and **do not include any overlapping responses**.

This is a **bad** question: it is two questions in one and the responses overlap.

Do you read and how many hours a week do you read?

☐ Never ☐ 0–1 hour ☐ 2–5 hours

☐ More than 6 hours

This is a **good** question.

Do you read books? ☐ Yes ☐ No   If your answer is yes, how many hours a week, on average, do you read books?

☐ Up to 2 hours ☐ Between 2 and 4 hours ☐ Over 4 hours

The data handling cycle

- There are **four** parts to the data handling cycle.
  - **Outlining** the problem. This usually involves stating a **hypothesis**, which is the idea being tested. For example, 'The girls did better in the maths test than the boys did' is a hypothesis.
  - Stating how the data will be **collected**, for example on a survey sheet or questionnaire.
  - Saying how the data will be **processed** and **represented**. This will usually involve working out means and ranges and showing the data in a suitable diagram such as a pie chart or a bar chart.
  - **Interpreting** the results and making a **conclusion**. This should be related to the original hypothesis or problem. For example, 'The hypothesis is correct as the mean for the girls is higher than the mean for the boys'.

REMEMBER

When you are asked to describe the data handling cycle, make sure you describe all four parts. Marks will be allocated for the overall quality and logic of your answer.

Questions

Grade D

1 Here are two questions used in a survey about recycling. Give a reason why each question is not a good one.

a Recycling is a waste of time and does not help the environment. Don't you agree?

☐ Yes ☐ No

b How many times a month do you use a bottle bank?

☐ Never ☐ 2 times or less

☐ More than 4 times

Grade C

AU 2 A gardener grows courgettes outside and in a greenhouse. These are the lengths of 10 courgettes grow outside:

12 14 11 15 17 9 16 12 11 18

Ten courgettes from the greenhouse also have their lengths measured. The mean length is 15.5 cm. The range of the length is 6 cm. Investigate the hypothesis: 'Courgettes grown in the greenhouse are bigger than those grown outside.'

Statistics grade booster

I can…

- [ ] draw and read information from bar charts, dual bar charts and pictograms
- [ ] find the mode and median of a list of data
- [ ] understand basic terms such as 'certain', 'impossible', 'likely'

You are working at **Grade G** level.

- [ ] work out the total frequency from a frequency table and compare data in bar charts
- [ ] find the range of a set of data
- [ ] find the mean of a set of data
- [ ] understand that the probability scale runs from 0 to 1
- [ ] calculate the probability of events with equally likely outcomes
- [ ] interpret a simple pie chart

You are working at **Grade F** level.

- [ ] read information from a stem-and-leaf diagram
- [ ] find the mode, median and range from a stem-and-leaf diagram
- [ ] list all the outcomes of two independent events and calculate probabilities from lists or tables
- [ ] calculate the probability of an event not happening when the probability of it happening is known
- [ ] draw a pie chart

You are working at **Grade E** level.

- [ ] draw an ordered stem-and-leaf diagram
- [ ] find the mean of a frequency table of discrete data
- [ ] find the mean from a stem-and-leaf diagram
- [ ] predict the expected number of outcomes of an event
- [ ] draw a line of best fit on a scatter diagram
- [ ] recognise the different types of correlation
- [ ] design a data collection sheet
- [ ] draw a frequency polygon for discrete data

You are working at **Grade D** level.

- [ ] find an estimate of the mean from a grouped table of continuous data
- [ ] draw a frequency diagram for continuous data
- [ ] calculate the relative frequency of an event from experimental data
- [ ] interpret a scatter diagram
- [ ] use a line of best fit to predict values
- [ ] design and criticise questions for questionnaires.

You are working at **Grade C** level.

Basic number

AQA 1/2/3 EDEXCEL 1/2/3 OCR 1/2/3

Times tables

- It is essential that you know the **times tables** up to 10 × 10.
- If you take away the 'easy' tables for 1, 2, 5 and 10 and then the 'repeated' tables such as 3 × 4 which is the same as 4 × 3, there are only 21 tables facts left to learn!

1	2	3	4	5	6	7	8	9	10
2	4	6	8	10	12	14	16	18	20
3	6	9	12	15	18	21	24	27	30
4	8	12	16	20	24	28	32	36	40
5	10	15	20	25	30	35	40	45	50
6	12	18	24	30	36	42	48	54	60
7	14	21	28	35	42	49	56	63	70
8	16	24	32	40	48	56	64	72	80
9	18	27	36	45	54	63	72	81	90
10	20	30	40	50	60	70	80	90	100

×	1	2	3	4	5	6	7	8	9	10
1										
2										
3			9							
4			12	16						
5										
6			18	24		36				
7			21	28		42	49			
8			24	32		48	56	64		
9			27	36		54	63	72	81	
10										

Order of operations and BODMAS

- When you are doing calculations, there is an order of operations that you must follow.

> 4 + 6 × 2 is often calculated as 20 because 4 + 6 = 10 and 10 × 2 = 20.
>
> This is wrong. Multiplication (×) must be done before addition (+), so the correct answer is 4 + 12 = 16.

> **REMEMBER**
> Always use brackets to make calculations clear.

- The order of operations is given by **BODMAS**, which stands for **B**rackets, **O**rder (or power), **D**ivision, **M**ultiplication, **A**ddition, **S**ubtraction.

- Brackets are used to show that parts of the calculation must be done first.

> (4 + 6) × 2 = 20 as the brackets mean 'work out 4 + 6 first'.

Questions

Grade G

1 a Write down the answer to each part.

   i 4 × 6   ii 3 × 7   iii 5 × 9

   iv 6 × 8   v 9 × 8   vi 7 × 7

  b Write down the answer to each part.

   i 32 ÷ 4   ii 35 ÷ 5   iii 63 ÷ 9

   iv 24 ÷ 8   v 81 ÷ 9   vi 56 ÷ 7

  c Write down the answer to each part. There is a remainder in each case.

   i 40 ÷ 7   ii 55 ÷ 6   iii 43 ÷ 6

   iv 28 ÷ 3   v 60 ÷ 9   vi 44 ÷ 7

Grade G

2 a Work out each of these.

   i 2 + 3 × 4   ii 10 − 2 × 2

   iii 12 + 6 ÷ 2   iv 15 − 8 ÷ 4

   v 4 × 8 ÷ 2   vi 20 ÷ 5 × 4

  b Work out each of these. Remember to work out the brackets first.

   i (5 + 4) × 3   ii (9 − 3) × 5

   iii (15 + 7) ÷ 2   iv (14 − 2) ÷ 4

   v (4 × 9) ÷ 6   vi 20 ÷ (5 × 4)

  c Put brackets into these calculations to make them true.

   i 5 × 6 + 1 = 35   ii 18 ÷ 2 + 1 = 6

   iii 25 − 10 ÷ 5 = 3   iv 20 + 12 ÷ 4 = 8

Number

Place value

- The **value** of a digit depends on its position or **place** in the number.
- The **place value** depends on the column heading above the digit.
- The first four column headings are **thousands**, **hundreds**, **tens** and **units**.

The number 4528 has 4 thousands, 5 hundreds, 2 tens and 8 units.

Thousands	Hundreds	Tens	Units
4	5	2	8

- To make them easier to read, the digits in numbers greater than 9999 are grouped in blocks of three.

Read 3 456 210 as 'three million, four hundred and fifty-six thousand, two hundred and ten' and 57 643 as 'fifty-seven thousand, six hundred and forty-three'.

Rounding

- Most numbers used in everyday life are **rounded**.

People may say, 'It takes me 30 minutes to drive to work,' or 'There were forty-two thousand people at the match last weekend.'

- Numbers can be rounded to the **nearest whole number**, the **nearest ten**, the **nearest hundred** and so on.

76 is rounded to 80 to the nearest 10, 235 is 200 to the nearest hundred and 240 to the nearest ten.

- The convention is that a **halfway** value rounds **upwards**.

2.5 is rounded to 3 to the nearest whole number.

REMEMBER

Show any 'carried' or 'borrowed' digits clearly.

Column addition and subtraction

- When adding or subtracting numbers without using a calculator, write the numbers in **columns**.
- Line up the **units digits**.
- **Start** adding or subtracting with the **units digits**.

```
    3 6 3 7
  +   7 4 8
    4 3 8 5
      1   1
```

```
        7  14  1
    2   8   5   4
  - 1   3   6   8
    1   4   8   6
```

Questions

Grade G

1 a In the number 3572 what is the value of the digit 5?

   b Write the number 27 708 in words.

   c Write the number 'Two million, four hundred and six thousand, five hundred and two' in numerals.

Grade G

2 a Round these numbers to the nearest ten.

   i 59   ii 142   iii 45

   b Round these numbers to the nearest hundred.

   i 682   ii 732   iii 1250

Grade F

3 Work out the following.

   a   2158   b   4215
      + 3672   − 1637

Number

Multiplying and dividing by single-digit numbers

- When **multiplying** numbers, without a calculator, write the numbers in **columns**.

  Write 4 × 57 as
  ```
      5 7
  ×     4
  ```

- **Start** multiplying with the **units digit** and show any 'carried' digits clearly.

  ```
      5 7
  ×     4
  ─────────
    2 2 8
      2
  ```

  REMEMBER
  Show any carried digits clearly.

- When **dividing** numbers, without a calculator, write it as a **short division**.

  Write 536 ÷ 8 as   8) 536

- **Start** dividing at the **left-hand side**.

  ```
        6 7
  8 ) 5 3 ⁵6
  ```

Problems in words

- When the **basic four operations** (**addition**, **subtraction**, **multiplication** and **division**) are used in real life, it is important to establish which operation to use.

  REMEMBER
  Write the necessary calculation in columns.

- The word **sum** means **addition**.

  'Find the sum of 123 and 45' means 123 + 45.

- The word **difference** means **subtract the smaller from the larger**.

  'Find the difference between 36 and 98' means 98 − 36.

- The word **product** means **multiply**.

  'Find the product of 8 and 26' means 8 × 26.

Questions

Grade F

1 a Work these out.
   i 7 × 39   ii 6 × 54
   b Work these out.
   i 208 ÷ 8   ii 238 ÷ 7
   c How many days are there in 12 weeks?
   d 180 plants are packed into trays of six plants. How many trays will there be?

Grade F

2 a Find the sum of 382 and 161.
   b Find the difference between 164 and 57.
   c Find the product of 8 and 62.
   d 364 fans travelled to a football match in seven coaches. Each coach was full and had the same number of seats. How many fans were there in each coach?

Grade C

PS 3 The sum of my age and my son's age is 50. Last year I was three times as old as my son. What is my age this year?

Fractions

AQA 1/2 EDEXCEL 2 OCR 2/3

Fractions of a shape

- A **fraction** is part of a whole.

- The number of parts the shape is divided into is called the **denominator** and is the bottom of the fraction.

- The number of parts required is called the **numerator** and is the top of the fraction.

> **REMEMBER**
> Remember to count up the shaded and unshaded parts to find the denominator.

This shape is divided into eight parts and five of them are shaded.

The shaded part is $\frac{5}{8}$ of the whole.

Adding and subtracting simple fractions

- If two fractions have the **same denominator** they can be added or subtracted.

- The denominator stays the same and the **numerators** are **added** or **subtracted**.

$$\frac{2}{7} + \frac{3}{7} = \frac{5}{7} \qquad \frac{8}{9} - \frac{4}{9} = \frac{4}{9}$$

Equivalent fractions

- Any fraction can be written in many different ways. These are called **equivalent fractions**.

$$\frac{1}{2} = \frac{2}{4} = \frac{3}{6} \cdots$$

- Equivalent fractions are two fractions that represent the **same part** of a **whole**.

- Any fraction has an **unlimited number** of equivalent fractions.

$$\frac{5}{8} = \frac{10}{16} = \frac{15}{24} = \frac{50}{80} = \frac{100}{160} \cdots$$

Questions

Grade G

1 What fraction is shaded in each of these diagrams?

a b

c d e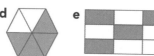

Grade F

3 a Which of the following fractions are equivalent to $\frac{1}{2}$?

$$\frac{3}{6} \qquad \frac{5}{7} \qquad \frac{3}{2} \qquad \frac{5}{10} \qquad \frac{8}{10} \qquad \frac{20}{40}$$

b Which of the following fractions are equivalent to $\frac{2}{5}$?

$$\frac{3}{6} \qquad \frac{4}{10} \qquad \frac{5}{20} \qquad \frac{8}{20} \qquad \frac{40}{100} \qquad \frac{1}{10}$$

c Which two diagrams in question **1** show equivalent fractions?

Grade G

2 Work these out.

a i $\frac{2}{11} + \frac{3}{11}$ **ii** $\frac{2}{5} + \frac{1}{5}$ **iii** $\frac{1}{8} + \frac{5}{8}$

b i $\frac{7}{9} - \frac{1}{9}$ **ii** $\frac{4}{5} - \frac{1}{5}$ **iii** $\frac{5}{13} - \frac{3}{13}$

Grade D

AU 4 Here are three fractions: $\frac{1}{2}$ $\frac{4}{8}$ $\frac{1}{8}$

Give a reason why each of them could be the odd one out.

Number

Equivalent fractions and cancelling

- A **base fraction** is one that is written in its **simplest terms**.

- A fraction is in its simplest terms if there is no number that is a factor of both the numerator and the denominator.

- Equivalent fractions can be found by **multiplying** both the **numerator** and the **denominator** by the **same number**.

$$\frac{3}{8} \times \frac{5}{5} = \frac{15}{40} \dots$$

- Equivalent fractions can be reduced to a base fraction by **cancelling**.

- To cancel a fraction, look for the **highest common factor** of the numerator and the denominator.

> **REMEMBER**
> Always cancel fractions as far as possible. It makes subsequent calculations easier.

To cancel $\frac{12}{15}$ look for the biggest number that divides into both 12 and 15.

This is 3. Divide both 12 and 15 by 3 to give $\frac{4}{5}$.

$$\frac{\overset{4}{\cancel{12}}}{\underset{5}{\cancel{15}}} = \frac{4}{5}$$

> **REMEMBER**
> Cross out the numbers when you cancel them.

Top-heavy fractions and mixed numbers

- A **top-heavy** fraction is a fraction in which the **numerator** is bigger than the **denominator**.

- Top-heavy fractions are also known as **improper** fractions.

- Fractions that are not top-heavy are called **proper** fractions.

- A **mixed number** is a mixture of a whole number and a proper fraction.

- To change a top-heavy fraction into a mixed number, **divide** the numerator by the denominator to find the whole number, then the **remainder** is the numerator of the proper fraction.

> **REMEMBER**
> You will need to convert mixed numbers to top-heavy fractions and vice versa to do addition, subtraction, multiplication and division problems.

$$\frac{13}{6} = 2\frac{1}{6} \text{ because } 13 \div 6 = 2 \text{ remainder } 1.$$

- To change a mixed number into a top-heavy fraction, **multiply** the whole number by the denominator and **add** the result to the numerator, to find the numerator of the top-heavy fraction.

$$3\frac{3}{4} = \frac{15}{4} \text{ because } 4 \times 3 + 3 = 15.$$

Questions

Grade G

1 a Find the numbers missing from the boxes.

   i $\frac{3}{5} \to \frac{\times 6}{\times 6} = \frac{\square}{30}$ **ii** $\frac{2}{3} \to \frac{\times 8}{\times 8} = \frac{16}{\square}$ **iii** $\frac{5}{8} = \frac{\square}{32}$

 b Cancel each of the following fractions.

   i $\frac{4}{10}$ **ii** $\frac{3}{15}$ **iii** $\frac{6}{20}$

   iv $\frac{15}{25}$ **v** $\frac{15}{21}$

 c Write down one equivalent fraction for each of these.

   i $\frac{3}{4}$ **ii** $\frac{1}{6}$ **iii** $\frac{3}{8}$

Grade F

2 a Change the following top-heavy fractions into mixed numbers.

   i $\frac{12}{5}$ **ii** $\frac{13}{4}$ **iii** $\frac{16}{7}$

   iv $\frac{21}{8}$ **v** $\frac{17}{3}$

 b Change the following mixed numbers into top-heavy fractions.

   i $2\frac{2}{3}$ **ii** $4\frac{3}{5}$ **iii** $2\frac{6}{7}$

   iv $2\frac{1}{4}$ **v** $3\frac{5}{8}$

Adding and subtracting fractions

- When two fractions are added (or subtracted) there are four possible outcomes.
  - The answer will be a proper fraction that does not need to be cancelled.

$$\frac{1}{7} + \frac{3}{7} = \frac{4}{7}$$

  - The answer will be a proper fraction that needs to be cancelled.

$$\frac{5}{9} - \frac{2}{9} = \frac{3}{9} = \frac{1}{3}$$

  - The answer will be a top-heavy fraction that does not need to be cancelled, so the fraction is converted to a mixed number.

$$\frac{3}{5} + \frac{4}{5} = \frac{7}{5} = 1\frac{2}{5}$$

REMEMBER

Always cancel before converting the top-heavy fraction into a mixed number.

  - The answer will be a top-heavy fraction that needs to be cancelled, then the fraction is converted to a mixed number.

$$\frac{7}{9} + \frac{5}{9} = \frac{12}{9} = \frac{4}{3} = 1\frac{1}{3}$$

Finding a fraction of a quantity

- To find a fraction of a quantity just multiply the fraction by the quantity.

To find $\frac{3}{5}$ of 25, $\frac{1}{5}$ of 25 = 5, so $\frac{3}{5} \times 25 = 3 \times 5 = 15$.

Find $\frac{2}{7}$ of 49 kg. $\frac{1}{7}$ of 49 = 7

$\frac{2}{7}$ of 49 = 2 × 7 = 14 g

- To compare fractions of numbers of quantities, work out the fractions first.

Which is the larger number, $\frac{2}{5}$ of 40 or $\frac{3}{7}$ of 35?

$\frac{1}{5}$ of 40 = 8

$\frac{2}{5}$ of 40 = 2 × 8 = 16

$\frac{1}{7}$ of 35 = 5

$\frac{3}{7}$ of 35 = 3 × 5 = 15   So $\frac{2}{5}$ of 40 is larger.

REMEMBER

Divide the quantity by the denominator to find the unit fraction of the quantity, then multiply this unit fraction by the numerator.

Questions

Grade G

1 Add or subtract the following fractions. Cancel the answers and/or make into mixed numbers if necessary.

a   i $\frac{1}{5} + \frac{3}{5}$ ii $\frac{5}{7} - \frac{3}{7}$ iii $\frac{1}{9} + \frac{4}{9}$ iv $\frac{2}{3} - \frac{1}{3}$

b   i $\frac{1}{6} + \frac{1}{6}$ ii $\frac{3}{10} - \frac{1}{10}$ iii $\frac{1}{9} + \frac{5}{9}$ iv $\frac{5}{12} - \frac{1}{12}$

c   i $\frac{6}{9} + \frac{7}{9}$ ii $\frac{5}{13} + \frac{10}{13}$ iii $\frac{7}{11} + \frac{8}{11}$ iv $\frac{2}{3} + \frac{2}{3}$

d   i $\frac{7}{8} + \frac{5}{8}$ ii $\frac{7}{10} + \frac{9}{10}$ iii $\frac{5}{9} + \frac{7}{9}$ iv $\frac{11}{12} + \frac{7}{12}$

Grade F

2 Calculate the following.

a $\frac{2}{5}$ of 20   b $\frac{5}{8}$ of 32

c $\frac{5}{6}$ of 30   d $\frac{3}{4}$ of £300

e $\frac{2}{9}$ of 81 kg   f $\frac{2}{3}$ of 6 hours

g Which is the larger number, $\frac{2}{3}$ of 21 or $\frac{3}{4}$ of 20?

h Which is the larger number, $\frac{4}{7}$ of 63 or $\frac{7}{8}$ of 40?

Number

Multiplying fractions

E

- To multiply two fractions, simply multiply the numerators to get the new numerator and multiply the denominators to get the new denominator.

$$\frac{3}{5} \times \frac{2}{7} = \frac{3 \times 2}{5 \times 7} = \frac{6}{35}$$

- If possible **cancel** numbers on the top and bottom before multiplying.

In $\frac{5}{6} \times \frac{9}{10}$ cancel 5 from 5 and 10 and cancel 3 from 6 and 9.

$$\frac{\cancel{5}^{1}}{\cancel{6}^{2}} \times \frac{\cancel{9}^{3}}{\cancel{10}^{2}} = \frac{3}{4}$$

> **REMEMBER**
> Always cancel before multiplying, as it makes the calculations easier and you will not have to cancel the final answer.

One quantity as a fraction of another

D

- To write one quantity as a fraction of another, write the **first quantity** as the **numerator** and the **second quantity** as the **denominator**.

What is £8 as a fraction of £20?

Write as $\frac{8}{20}$ then cancel to $\frac{2}{5}$.

Problems in words

D

- In examinations, most fraction problems will be set out as real-life problems expressed in words.

- Decide on the calculation, write it down then work it out.

John eats a quarter of a cake and Mary eats half of what is left. What fraction of the original cake is left?

John eats $\frac{1}{4}$. Mary eats $\frac{1}{2} \times \frac{3}{4} = \frac{3}{8}$.

So $\frac{3}{8}$ of the cake is left.

> **REMEMBER**
> Examination questions often ask for the fraction to be given in its simplest form. This means it has to be cancelled down.

Questions

Grade E

1 Multiply the following fractions. Give the answers in their simplest form.

a $\frac{1}{3} \times \frac{3}{5}$ b $\frac{5}{8} \times \frac{3}{5}$ c $\frac{1}{2} \times \frac{4}{9}$

d $\frac{2}{3} \times \frac{9}{10}$ e $\frac{2}{9} \times \frac{1}{8}$ f $\frac{3}{8} \times \frac{4}{15}$

g $\frac{4}{9} \times \frac{3}{8}$ h $\frac{7}{8} \times \frac{10}{21}$

Grade D

2 a What fraction of 25 is 10?

b In a class of 28 students, 21 are right-handed. What fraction is this?

c What fraction is 20 minutes of an hour?

Grade D

3 a In a mixed box of tapes and CDs, $\frac{2}{3}$ of the items were CDs. What fraction were tapes?

b In a packet of toffees $\frac{1}{2}$ are plain toffees, $\frac{3}{8}$ are nut toffees and the rest are treacle toffees. What fraction are treacle toffees?

c Ken earns £400 a week. One week he earns a bonus of $\frac{1}{5}$ of his wages. How much does he earn that week?

Grade D

PS 4 Jim is 40. His daughter Jane is $\frac{2}{5}$ of his age. His son John is $\frac{7}{20}$ of his age.

How much older is Jane than John?

Rational numbers

AQA 1/2   EDEXCEL 2/3   OCR 2/3

Converting fractions into decimals

- As the line separating the top and bottom of a fraction (known as the vinculum) means divide, all you have to do to convert a fraction to a decimal is divide the numerator by the denominator.

$$\frac{3}{5} = 3 \div 5 = 0.6, \frac{5}{8} = 5 \div 8 = 0.625, \frac{9}{25} = 9 \div 25 = 0.36, \frac{2}{3} = 2 \div 3 = 0.6666...$$

Rational numbers

- A **rational number** is any number that can be expressed as a **fraction**.
- Some fractions result in **terminating decimals**.

$$\frac{1}{16} = 0.0625, \frac{3}{64} = 0.046\,875, \frac{7}{40} = 0.175,$$

$$\frac{19}{1000} = 0.019, \text{ so all give terminating decimals.}$$

> **REMEMBER**
> The only fractions that give terminating decimals are those with a denominator that is a power of 2, 5 or 10, or a combination of these.

- Some fractions result in **recurring** decimals.

$$\frac{1}{3} = 0.3333...$$

- **Recurrence** is shown by dots over the recurring digit or digits.

0.3333... becomes $0.\dot{3}$ in recurrence or dot notation.

0.277 777... becomes $0.2\dot{7}$ 0.518 518 518... becomes $0.\dot{5}1\dot{8}$

- To convert a fraction into a decimal, just divide the numerator by the denominator.

$$\frac{7}{20} = 7 \div 20 = 0.35, \frac{5}{11} = 5 \div 11 = 0.454\,545... = 0.\dot{4}\dot{5}$$

Converting terminating decimals into fractions

- Depending on the number of places in the decimal, the denominator will be 10, 100, 1000, ...

$$0.7 = \frac{7}{10}, 0.036 = \frac{36}{1000} = \frac{9}{250}, 2.56 = \frac{256}{100} = \frac{64}{25} = 2\frac{14}{25}$$

> **REMEMBER**
> Count the number of decimal places. This is the same as the number of zeros in the denominator.

Questions

Grade E

1 Write the following fractions as decimals.

   a $\frac{3}{8}$ b $\frac{2}{5}$ c $\frac{9}{20}$

   d $\frac{2}{9}$ e $\frac{17}{32}$ f $\frac{5}{11}$

Grade E

2 a Put the following fractions in order, starting with smallest.

    $\frac{13}{20}$ $\frac{2}{3}$ $\frac{16}{25}$ $\frac{7}{11}$

  b Put the following fractions in order, starting with the largest.

    $\frac{9}{20}$ $\frac{4}{9}$ $\frac{5}{11}$ $\frac{12}{25}$

Grade C

3 Use recurrence or dot notation to write each of these.

   a 0.363 636...   b 0.615 615 615...

   c 0.3666...   d $\frac{2}{3}$

   e $\frac{1}{6}$ f $\frac{7}{9}$

Grade C

4 Convert the following terminating decimals into fractions. Cancel your answers if possible.

   a 0.8   b 0.65   c 0.125

   d 2.45   e 0.025   f 0.888

Negative numbers

Negative numbers

- The **natural numbers** are 1, 2, 3, 4, 5, …
- The **counting numbers** are 0, 1, 2, 3, 4, 5, …
- **Negative numbers** are those numbers below zero such as –1, –2, –8.3.
- Negative numbers are used in everyday life to give **temperatures**, **distances below** certain points (such as sea level) and to indicate **overdrawn bank balances**, for example.
- Negative numbers are also sometimes known as **directed numbers** when they are used with a number line.

> **REMEMBER**
> The larger the actual number in a negative number, the smaller it is. i.e. –9 is smaller than –3.

The number line

- Negative numbers can be represented on a **number line**.
- Number lines can be **horizontal** or **vertical**.

```
 –7   –6   –5   –4   –3   –2   –1    0    1    2    3    4    5    6    7
            negative                          positive
```

Vertical number line: 7, 6, 5, 4, 3, 2, 1, 0, –1, –2, –3, –4, –5, –6, –7

- Negative numbers are to the **left** or **below** zero.
- Positive numbers are to the **right** or **above** zero.
- The further to the **left** or the further **down** the number line, the **smaller** the number.

Questions

Grade G

1 Complete the following sentences.

a If +£7 means a profit of seven pounds, then … means a loss of seven pounds.

b If +300 m means 300 metres above sea level, then … means 40 metres below sea level.

c If –6 means 6 hours before midnight, then 8am the next day would be represented by … .

d If 40 °F is 8 degrees above freezing in the Fahrenheit scale, then 8 degrees below freezing is given by … °F.

e If –100 km means 100 kilometres south of Leeds, then 50 km north of Leeds would be represented by … .

f If –100 km means 100 kilometres west of Leeds, then 2 km east of Leeds would be represented by … .

Grade F

2 Use the number line above to answer this question.

a Put < or > in each statement to make it true.

   i –3 … 3   ii –8 … –10   iii –6 … –2

b Write down the next two terms of this sequence.

   11, 8, 5, 2, … , …

c For each pair of numbers, write down the number that is halfway between them.

   i 2 and 8   ii –8 and 2   iii –6 and –1

Grade F

AU 3

More than –9   Less than –2

Divides exactly by 3

What numbers are being described?

Addition and subtraction with negative numbers

- Adding a positive number is a move to the right.

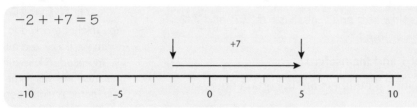

$-2 + +7 = 5$

- Subtracting a positive number is a move to the left.

$+4 - +8 = -4$

- Adding a negative number is a move to the left.

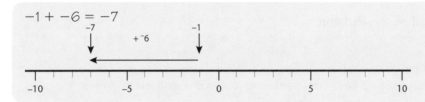

$-1 + -6 = -7$

- Subtracting a negative number is a move to the right.

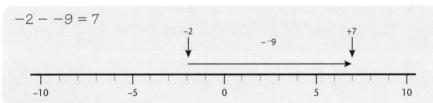

$-2 - -9 = 7$

> **REMEMBER**
>
> An easy way to remember these rules is: If the signs are the same the result is a plus, if the signs are different it is a minus.
>
> + (+) = +   – (–) = +
> + (–) = –   – (+) = –
>
> –3 + –2 is the same as
> – 3 – 2 = – 5
>
> 2 – –8 is the same as
> 2 + 8 = 10

Questions

1 Work out the following. Use a number line to help if necessary.

a 6 − 10	**b** −3 − 9	**c** −6 + 8
d +5 − 8	**e** −9 + 9	**f** 6 − 8 − 2
g 3 − +9	**h** −2 − −7	**i** −3 − +8 − +3
j −9 − −5	**k** 7 − −5	**l** −7 + 8 − 9
m 4 − 6 + 9	**n** −5 − 8 − 9	**o** −6 + −8 + +6

2 a What is the sum of +3, −5 and −8?

b What is the sum of −6, −7 and −11?

c What is the difference between −10 and +11?

d What is the difference between −8 and −14?

3 Write down the calculation shown on each number line.

a

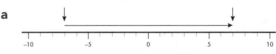

b

c

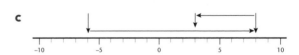

AU 4 a Write two **positive** numbers in the boxes to make the equation true.

$\square - \square = -2$

b Write two **negative** numbers in the boxes to make the equation true.

$\square - \square = -2$

More about number

Multiples

- A **multiple** of a number is any number in the **times table**.

  > $5 \times 7 = 35$, so 35 is a multiple of 5 and 7 as it is in the 5 and 7 times tables. 35 is also a multiple of 1 and 35.

- All numbers are multiples of **1** and **themselves**.

- There are some rules for spotting if numbers are in certain times tables.

  - **Multiples of 2** (which are also even numbers) always end in 0, 2, 4, 6 or 8.

  - **Multiples of 3** have digits that add up to a multiple of 3.

    > 372 is a multiple of 3 because $3 + 7 + 2 = 12$ which is 4×3.

  - **Multiples of 5** always end in 0 or 5.

  - **Multiples of 9** have digits that add up to a multiple of 9.

    > 738 is a multiple of 9 because $7 + 3 + 8 = 18 = 2 \times 9$.

  - **Multiples of 10** always end in 0.

- Other multiples can be tested on a **calculator**.

  > To test 196:
  > As it ends in 6 it is a multiple of 2.
  > $1 + 9 + 6 = 16$, which is not in the 3 or 9 times table so it is not a multiple of 3 or 9.
  > It does not end in 0 or 5 so it is not a multiple of 5 or 10.
  > $196 \div 4 = 49$, so as the answer is a whole number it must be a multiple of 4.
  > $196 \div 7 = 28$, so as the answer is a whole number it must be a multiple of 7.
  > $196 \div 8 = 24.5$, so as the answer is not a whole number it is not a multiple of 8.
  > So 196 is a multiple of 2, 4 and 7.

> **REMEMBER**
> You might be asked to say, for example, if 729 is a multiple of 9. You need to know the rules because this will come on the non-calculator paper.

> **REMEMBER**
> This is why you need to know your tables!

Factors

- A **factor** of a number is any whole number that divides into it exactly.

  > The factors of 20 are {1, 2, 4, 5, 10, 20}.

- **1** is always a factor and so is the **number itself**.

- When you find one factor there is always **another** that goes with it, unless you are investigating a square number.

  > $1 \times 12 = 12$, $2 \times 6 = 12$, $3 \times 4 = 12$ so the factors of 12 are {1, 2, 3, 4, 6, 12}.

  > $1 \times 16 = 16$, $2 \times 8 = 16$, $4 \times 4 = 16$, so the factors of 16 are {1, 2, 4, 8, 16}.

> **REMEMBER**
> On the non-calculator paper, you may be asked to find the factors of numbers up to 60. Any bigger numbers will be on the calculator paper.

Questions

Grade F

1 a Write down the first five multiples of each number.

  i 6   **ii** 13   **iii** 25

b From the list at the top of the next column, identify those numbers that are multiples of:

  i 2   **ii** 3   **iii** 5
  iv 9   **v** 10.

| 121 | 250 | 62 | 78 | 90 | 85 | 108 |
| 144 | 35 | 81 | 96 | 120 | 333 | 125 |

c Is 819 a multiple of 9?

Grade F

2 a Write down the factors of each number.

  i 24   **ii** 15   **iii** 50   **iv** 40

b Use a calculator to find the factors of 144.

Primes and squares

Prime numbers

- A **prime number** is a number with only two **factors**.

  17 has only the factors 1 and 17, 3 has only the factors 1 and 3.

- The factors of a prime number are always **1** and the number **itself**.

- There is no rule or pattern for spotting prime numbers, you will have to **learn them**.

  The prime numbers up to 50 are:

  2, 3, 5, 7, 11, 13, 17, 19, 23, 29, 31, 37, 41, 43 and 47.

- All prime numbers are **odd**, **except for 2**, which is the **only even** prime number.

> **REMEMBER**
> 1 is *not* a prime number as it only has one factor (1).

Square numbers

- The sequence 1, 4, 9, 16, 25, 36, ... is called the **sequence of square numbers**.

- The square numbers can be represented by **square patterns**.

> **REMEMBER**
> You are expected to know the square numbers up to 15×15 (= 225).

1×1 2×2 3×3 4×4 5×5

- Square numbers can be calculated as 1×1, 2×2, 3×3 and so on.

- 1×1, 2×2, 3×3 ... can be written as numbers to the power 2: 1^2, 2^2, 3^2, ...

- The power 2 is referred to as the number **squared**.

  Read 7^2 as '7 squared'.

Questions

Grade E

1 a Write down the factors of each number.

  i 18   **ii** 19   **iii** 20

  b Which of the numbers in part (a) is a prime number?

  c From the list below, identify those that are prime numbers.

21	25	61	79	9
83	17	41	35	81
29	1	33	15	

  d What is the only even prime number?

Grade E

AU 2 a Continue the following sequence to 10 terms.

  1, 4, 9, 16, 25, ... , ... , ... , ... , ...

  b Work out the value of each number.

  i 11^2 **ii** 12^2 **iii** 13^2

  iv 14^2 **v** 15^2

  c Continue the following number pattern for three more lines.

  $$1 + 3 = 4$$
  $$1 + 3 + 5 = 9$$
  $$1 + 3 + 5 + 7 = 16$$

Roots and powers

Square roots

- The **square root** of a given number is a number that, when multiplied by itself, produces the given number.

  > The square root of 16 is 4 as $4 \times 4 = 16$.

- Square roots can also be **negative**.

  > -4×-4 also equals 16.

- A square root is represented by the **symbol** $\sqrt{\ }$.

  > $\sqrt{25} = 5$

- Calculators have a **'square root' button**.

- Taking a square root is the **inverse operation** to squaring.

> **REMEMBER**
> If you are asked for $\sqrt{81}$, a single answer of 9 is correct but if you are asked for values of x that make $x^2 = 36$, then you should give answers of +6 and –6. One clue will be how many marks the question is worth. 2 marks means two answers.

Powers

- **Powers** are a convenient way of writing **repeated multiplications**.

  > $3 \times 3 \times 3 \times 3 \times 3 \times 3 = 3^6$, which is read as '3 to the power 6'.

- Powers are also called **indices** (singular **index**).

- Most calculators have **power buttons**. x^2 x^3 x^y

- The **power 2** has a special name: **'squared'**.

- The **power 3** has a special name: **'cubed'**.

- The **inverse operation** of cubing is taking the **cube root**, which is represented by the symbol $\sqrt[3]{\ }$.

  > $\sqrt[3]{8} = 2, \sqrt[3]{27} = 3$ $\sqrt[3]{\ }$

> **REMEMBER**
> You are expected to know the cube numbers 1^3, 2^3, 3^3, 4^3, 5^3 and 10^3.

Questions

Grade F

1 a Write down the positive square root of each of these numbers.
    i 81   **ii** 64   **iii** 25

  b For each statement, write down two values of x that make it true.
    i $x^2 = 4$ **ii** $x^2 = 16$ **iii** $x^2 = 100$

  c Use a calculator to work out these square roots.
    i $\sqrt{576}$ **ii** $\sqrt{6.25}$ **iii** $\sqrt{37.21}$

  d Write down the value of each of these numbers.
    i 5^3 **ii** 1^3 **iii** 10^3

  e Use your calculator to work out the value of each of these numbers.
    i $\sqrt[3]{1.728}$ **ii** $\sqrt[3]{4096}$ **iii** $\sqrt[3]{0.027}$

Grade D

2 a Write down the values of these numbers.
    i 3^3 **ii** 4^3 **iii** 10^3

  b Write these numbers, using power notation.
    i $4 \times 4 \times 4 \times 4 \times 4$
    ii $6 \times 6 \times 6 \times 6 \times 6 \times 6$
    iii $10 \times 10 \times 10 \times 10$
    iv $2 \times 2 \times 2 \times 2 \times 2 \times 2 \times 2$

  c Use a calculator to work out the values of the powers in part (**b**).

  d Continue the sequence of the powers of 2 up to 10 terms.
    2, 4, 8, 16, 32, … , … , … , … , …

Powers of 10

Powers of 10

> **REMEMBER**
> The positive power of 10 and the number of zeros after the number are the same, for example, $10^5 = 100\,000$, and 1 million is 10^6.

- The **decimal** system, which is the basis of our number system, is based on **powers of 10**.

- **Column headings** such as hundreds, tens and units are **powers of 10**.

The number 346.32 could be written as:

100	10	1		$\frac{1}{10}$	$\frac{1}{100}$
3	4	6	•	3	2

or

10^2	10^1	10^0		10^{-1}	10^{-2}
3	4	6	•	3	2

- The decimal point separates the whole numbers from the decimal fractions.

- The column headings after the decimal point are **tenths**, **hundredths** or 10^{-1}, 10^{-2}.

Multiplying and dividing by powers of 10

> **REMEMBER**
> Although, strictly speaking, the digits move, in reality it is easier to think of it as moving the decimal point.

- When you **multiply** by a power of 10, the digits of the number move to the **left**.

- When you **divide** by a power of 10, the digits move to the **right**.

$34.5 \times 10 = 345$

100	10	1		$\frac{1}{10}$
	3	4	•	5
3	4	5	•	0

$\times 10$

$45.9 \div 10^2 = 0.459$

10	1		$\frac{1}{10}$	$\frac{1}{100}$	$\frac{1}{1000}$
4	5	•	9		
		•	4	5	9

$\div 10^2$

- The number of **places the digits move** depends on the number of **zeros** or the power of 10.

Multiplying and dividing multiples of powers of 10

- When you **multiply** together two multiples of powers of 10, just multiply the non-zero digits and write the **total** of the zeros in both numbers at the end.

$200 \times 4000 = 800\,000, 500 \times 60 = 30\,000$

- When you **divide** one multiple of powers of 10 by another, just divide the non-zero digits and write the **difference** in the zeros in both numbers at the end.

$8000 \div 20 = 400, 20\,000 \div 40 = 500$

Questions

Grade E

1 Write down the answers.

  a 8×100 **b** 6.4×10

  c 2.5×10^2 **d** 0.3×10^3

  e 7.6×100 **f** 3.25×1000

  g $64 \div 10$ **h** $2.8 \div 100$

  i $390 \div 10^2$ **j** $0.75 \div 10$

  k $34 \div 10^3$ **l** $9.4 \div 10$

Grade E

2 Write down the answers.

  a 3000×200 **b** 40×300

  c 2000×70 **d** $4000 \div 20$

  e $6000 \div 300$ **f** $40\,000 \div 800$

Number

Prime factors

Prime factors

- When a number is written as a product of **prime factors** it is written as a multiplication consisting only of prime numbers.

    $30 = 2 \times 3 \times 5, 50 = 2 \times 5 \times 5 \text{ or } 2 \times 5^2$

- There are two ways to find prime factors: the **division method** and the **tree method**.

> **REMEMBER**
> If a number is even then 2 is an obvious choice as a divisor or part of the product, then look at 3, 5, 7, …

The division method to find the prime factors of 24

Divide by prime numbers until the answer is a prime number.

$$
\begin{array}{r}
2\;)\;2\;\;4 \\
2\;)\;1\;\;2 \\
2\;)\;\;\;\;6 \\
\hline
3
\end{array}
$$

So $24 = 2 \times 2 \times 2 \times 3$

The tree method to find the prime factors of 76

Keep splitting numbers into products until there are prime numbers at the ends of all the branches.

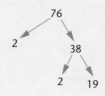

So $76 = 2 \times 2 \times 19$

- Products of prime factors can be expressed in **index form**.

    $24 = 2 \times 2 \times 2 \times 3 = 2^3 \times 3$

    $50 = 2 \times 5 \times 5 = 2 \times 5^2$

    $76 = 2 \times 2 \times 19 = 2^2 \times 19$

Questions

Grade D

1 a Write the following in index form.

   i $2 \times 2 \times 2 \times 3 \times 3$

   ii $2 \times 2 \times 3 \times 5 \times 5 \times 5$

   iii $3 \times 3 \times 5 \times 5$ **iv** $2 \times 3 \times 3 \times 3 \times 7$

 b Work out the values of the prime factor products in part (**a**).

Grade C

2 What numbers are represented by these products of prime factors?

  a $2 \times 3 \times 5$ **b** $2 \times 2 \times 3 \times 7$

  c $2 \times 5 \times 13$ **d** $2^2 \times 3^2$

  e $2^3 \times 5$ **f** $2^2 \times 3 \times 5^2$

Grade C

3 Use the division method to find the product of prime factors for each of these numbers. Give your answers in index form if possible.

  a 20   **b** 45   **c** 64   **d** 120

Grade C

4 Use the tree method to find the product of prime factors for each of these numbers. Give your answers in index form if possible.

  a 16   **b** 42   **c** 70   **d** 200

LCM and HCF

AQA 2   EDEXCEL 2   OCR 1

Lowest common multiple

Use the list method, as this is much easier and gets full credit, even if you have already done part of the prime factor method in an earlier part of the question.

- The **lowest common multiple (LCM)** of two numbers is the smallest number that appears in the times tables of both of the two numbers.

  The LCM of 6 and 7 is 42, the LCM of 8 and 20 is 40.

- There are two ways to find the LCM: the **prime factor method** and the **list method**.

The prime factor method

Find the LCM of 24 and 40.

Write 24 and 40 as products of prime factors: $24 = 2^3 \times 3$, $40 = 2^3 \times 5$

Now find the smallest product of prime factors that includes all the prime factors of 24 and 40.

This is $2^3 \times 3 \times 5 = 120$.

So the LCM of 24 and 40 is 120.

The list method

Find the LCM of 16 and 20.

Write out the 16 and 20 times tables, continuing until there is number that appears in both lists (a common multiple).

16 times table: 16, 32, 48, 64, (80), 96, 112, 128, …

20 times table: 20, 40, 60, (80), 100, 120, …

So the LCM of 16 and 20 is 80.

Highest common factor

- The **highest common factor (HCF)** of two numbers is the **biggest** number that **divides exactly** into the two numbers.

  The HCF of 16 and 20 is 4, the HCF of 18 and 42 is 6.

- There are two ways to find the HCF: the **prime factor method** and the **list method**.

The prime factor method

Find the HCF of 45 and 108.

Write 45 and 108 as products of prime factors: $45 = 3^2 \times 5$, $108 = 2^2 \times 3^3$

Now find the biggest product of prime factors that is included in the prime factors of 45 and 108.

This is $3^2 = 9$.

So the HCF of 45 and 108 is 9.

The list method

Find the HCF of 36 and 90.

Write out the factors of 36 and 90 then pick out the biggest number that appears in both lists.

Factors of 36: {1, 2, 3, 4, 6, 9, 12, (18), 36}

Factors of 90: {1, 2, 3, 5, 6, 9, 10, 15, (18), 30, 45, 90}

So the HCF of 36 and 90 is 18.

Questions

Grade C

1 a Find the LCM of each pair of numbers.
  i 5 and 6   ii 3 and 7   iii 3 and 13
  b Describe a connection between the LCM and the original numbers in part (a).
  c Find the LCM of each pair of numbers.
  i 6 and 9   ii 8 and 20   iii 15 and 25

Grade C

2 Find the HCF of the numbers in each pair.
  a 12 and 30   b 18 and 40   c 15 and 50
  d 16 and 80   e 24 and 60   f 12 and 25

Grade C

3 Find the LCM of the numbers in each pair.
  a 12 and 30   b 15 and 50   c 24 and 60

Powers

Multiplying powers

- When two powers of the **same base number** are **multiplied** together, then the new power is the **sum** of the original powers.

  $4 \times 8 = 32$, but $4 \times 8 = 2^2 \times 2^3 = 2^5 = 32$

Dividing powers

- When two powers of the **same base number** are **divided**, then the new power is the **difference** of the original powers.

  $243 \div 9 = 27$, but $243 \div 9 = 3^5 \div 3^2 = 3^3 = 27$

- Any number to the power 1 is just the number itself.

  $32 \div 16 = 2$, but $32 \div 16 = 2^5 \div 2^4 = 2^{5-4} = 2^1$

- Any number to the power 0 is always 1.

  $27 \div 27 = 1$, but $27 \div 27 = 3^3 \div 3^3 = 3^{3-3} = 3^0$

> **REMEMBER**
> You are often asked to write an expression such as $5^2 \times 5^3$ as a single power of 5, which would be 5^5. You do not have to work out the actual value unless you are asked to do so, and this would not be on the non-calculator paper anyway.

Multiplying and dividing powers with letters

- When you are working with **numbers**, using the **rules for multiplying and dividing powers** helps to simplify calculations but they can always be evaluated as numerical answers.

- When you use algebraic unknowns, or letters, as the base the **expression cannot be simplified**.

- Algebraic powers follow exactly the same rules as above.

  $x^2 \times x^6 = x^8$ $x^9 \div x^6 = x^3$

> **REMEMBER**
> Don't get confused with ordinary numbers when powers are involved. The rules above only apply to powers, so $2x^3 \times 3x^3 = 2 \times 3 \times x^3 \times x^2$ = $6x^5$ *not* $5x^5$.

Questions

Grade C

1 a Write each of the following as a single power of 2.

   i $2^3 \times 2^4$ **ii** $2^4 \times 2^5$ **iii** $2^3 \times 2^3$

  b Write each of the following as a single power of 3.

   i $3^5 \div 3^2$ **ii** $3^8 \div 3^4$ **iii** $3^6 \div 3^3$

  c Write each of the following as a single power of x.

   i $x^6 \times x^3$ **ii** $x^5 \times x^4$ **iii** $x^6 \times x^5$

  d Write each of the following as a single power of x.

   i $x^7 \div x^3$ **ii** $x^9 \div x^3$ **iii** $x^5 \div x^2$

  e Which of the following is the algebraic rule for $x^n \times x^m$?

   i x^{nm} **ii** x^{n+m} **iii** $(m+n)x$

  f Which of the following is the algebraic rule for $x^n \div x^m$?

   i x^{n-m} **ii** $x^{n \div m}$ **iii** $(m-n)x$

Grade C

2 Write down the value of each number.

  a 4^0 **b** 7^1

  c $8^5 \div 8^5$ **d** $6^4 \div 6^3$

Grade C

AU 3 a Which of the following is the correct answer to $2x^2 \times 5x^5$?

   i $7x^{10}$ **ii** $10x^7$ **iii** $7x^7$ **iv** $10x^7$

  b Which of the following is the correct answer to $12x^6 \div 3x^2$?

   i $9x^3$ **ii** $4x^3$ **iii** $4x^4$ **iv** $9x^4$

Long multiplication

- **Long multiplication** is used to multiply two numbers that both have more than one digit.
- There are several methods for long multiplication. The **four most common** are:
    - **Napier's bones** or **Chinese multiplication**
    - the **column** method
    - the **expanded column** method
    - the **box** method.

> **REMEMBER**
> In the GCSE non-calculator paper, you are only expected to multiply a two-digit number by a three-digit number at the most. Any calculations involving bigger numbers will be on the calculator paper.

Napier's bones or Chinese multiplication

Work out 27 × 42.

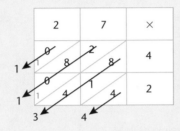

So 27 × 42 = 1134.
Note the carried digits.

The column method

Work out 36 × 28.

```
      3 6
  ×   2 8
  ─────────
    2 8 8
        4
    7 2 0
      1
  ─────────
  1 0 0 8
    1 1
```

> **REMEMBER**
> This is the traditional method but it can be confusing, with all the carried digits and the need to remember to put down a zero.

So 36 × 48 = 1008.

Expanded column method

Work out 47 × 52.

```
      5 2          5 2          2 0 8 0
  ×   4 0      ×     7      +      3 6 4
  ─────────     ─────────      ─────────
  2 0 8 0        3 6 4          2 4 4 4
                    1              1
```

So 47 × 52 = 2444.

> **REMEMBER**
> This method makes the calculation into two short multiplications.

The box method

Work out 23 × 45.

×	20	3
40	800	120
5	100	15

```
      8 0 0
      1 2 0
      1 0 0
  +     1 5
  ─────────
  1 0 3 5
      1
```

So 23 × 45 = 1035.

> **REMEMBER**
> Decide which method you prefer and stick to it. It doesn't matter which method you use in the GCSE as long as you get the answer right.

Questions

1 Use whichever method you prefer to work these out.

a 24 × 23	**b** 61 × 52	**c** 63 × 31
d 78 × 34	**e** 147 × 43	**f** 265 × 26

Long division

- Long division is used to divide a number with three or more digits by a number with two or more digits.

- There are two methods for long division:
  - the **traditional** or **Italian** method
  - the **repeated subtraction** or **chunking** method.

The traditional or Italian method

Work out $864 \div 24$.

```
        3 6
24 ) 8 6 4
     7 2
     ‾‾‾‾‾
     1 4 4
     1 4 4
     ‾‾‾‾‾
         0
```

1 How many 24s in 86?
There are 3 and $3 \times 24 = 72$.

2 $86 - 72 = 14$ bring the 4 down to make 144.

3 How many 24s in 144? There are 6 exactly.

This is basically a way of showing a 'short' division neatly.

```
        3 6
24 ) 8 ⁸6¹⁴4        So 864 ÷ 24 = 36.
```

Repeated subtraction or chunking

Work out $1564 \div 34$.

1 Start by writing down the easier multiples of 34.

$1 \times 34 = 34, 2 \times 34 = 68, 10 \times 34 = 340, 20 \times 34 = 680$

2 Keep on subtracting multiples of 34, taking away the biggest you can each time.

```
  1 5 6 4
    6 8 0      20 ×
  ‾‾‾‾‾‾‾
    8 8 4
    6 8 0      20 ×
  ‾‾‾‾‾‾‾
    2 0 4
      6 8      2 ×
  ‾‾‾‾‾‾‾
    1 3 6
      6 8      2 ×
  ‾‾‾‾‾‾‾
      6 8
      6 8      2 ×
  ‾‾‾‾‾‾‾
        0      46 ×
```

3 When the answer is 0 or less than 34, add up how many have been taken away.
So $1564 \div 34 = 46$.

Questions

Grade F

1 Use whichever method you prefer to work these out.

a $572 \div 22$ b $528 \div 33$ c $988 \div 26$
d $704 \div 16$ e $672 \div 24$ f $637 \div 13$

Real-life problems

- In a GCSE exam, long multiplication and long division are usually given in the context of **real-life problems**.

- You will need to **identify** the problem as a **multiplication** or **division** and then work it out by your preferred method.

- In the GCSE non-calculator paper, questions often ask such things as 'How many coaches are needed?' and the calculation gives a remainder. Remember that an extra coach will be needed to carry the remaining passengers. You cannot have a 'bit of a coach'.

A café uses 950 eggs per day. The eggs come in trays of 24. How many trays of eggs will the café need?

This is a division problem. Use the traditional method.

```
        3 9
24 ) 9 5 0
     7 2
     ───
     2 3 0
     2 1 6
     ─────
         1 4
```

The answer to the calculation $950 \div 24$ is 39 remainder 14.

The café will need 40 trays and will have 10 eggs left over.

> **REMEMBER**
> Show your working clearly because even if you make a small arithmetical error you will still get marks for method.

Decimal places

- The number of decimal places in a number is just the **number of digits after the decimal point**.

  2.34 has two decimal places (2 dp), 3.068 has three decimal places (3 dp).

- You need to be able to round numbers to one, two or three decimal places (1 dp, 2 dp, 3 dp).

- To round a number to one decimal place, look at the digit in the **second** decimal place. If it is less than 5 remove the unwanted digits. If it is 5 or more add 1 on to the digit in the **first** decimal place.

- Use the same method to round to two or three decimal places.

  3.5629 is 3.6 to 1 dp, 3.56 to 2 dp and 3.563 to 3 dp.

> **REMEMBER**
> There will always be at least one question in your GCSE that asks you to round a number.

Questions

Grade F

1 **a** There are 945 students in a school.
    There are 27 students in each tutor group.
    How many tutor groups are there?

   b To raise money for charity one tutor group decides that all 27 members will donate the cost of a KitKat. If a KitKat costs 42p, how much money do they raise?

Grade F

2 **a** How many decimal places are there in each of these numbers?
    i 4.09   **ii** 32.609   **iii** 4.5

   b Round these numbers to one decimal place.
    i 2.32   **ii** 6.08   **iii** 15.856

   c Round these numbers to two decimal places.
    i 3.454   **ii** 16.089   **iii** 7.625

   d Round these numbers to three decimal places.
    i 4.9743   **ii** 6.2159   **iii** 0.0076

Decimals

AQA 1/2 EDEXCEL 1/2/3 OCR 1/2/3

Adding and subtracting decimals

- When **adding** and **subtracting** decimals, you must use a **column** method.

- **Line up** the **decimal points** and use **zeros** to fill in any blanks.

- Do the addition or subtraction as normal, starting with the column on the right.

- The decimal point in the answer will be placed directly underneath the other decimal points.

Write 3.4 + 2.56 as:

```
  3 . 4 0
+ 2 . 5 6
```

REMEMBER
Make sure the decimal points are lined up.

```
  3 . 4 0          6 . ⁶7̷ ¹0
+ 2 . 5 6        - 2 . 4 8
─────────        ─────────
  5 . 9 6          4 . 2 2
```

Multiplying and dividing decimals by single-digit numbers

- Solve these problems in the same way as for normal short multiplication and division.

- As in addition and subtraction, the decimal point in the answer will be underneath the decimal point in the calculation.

Write 4.32 × 3 as:
```
      4 . 3 2
    ×       3
  ───────────
  1 2 . 9 6
```

and 3.45 ÷ 3 as:
```
        1 . 1 5
    3 ) 3 . 4 ¹5
```

Long multiplication with decimals

- Solve these problems in the same way as for **normal long multiplication**.

- As before, keep the decimal points **in line**.

3.26 × 24
```
      3 . 2 6
    ×     2 4
  ───────────
  1 3   0 4
  6 5   2 0
  ───────────
  7 8 . 2 4
```

Multiplying decimals

- Solve these problems in the same way as for **normal multiplications** but ignore the decimal points in the working.

- The number of decimal places in the answer will be the same as the **total** of **decimal places** in the numbers in the original calculation.

There are three decimal places in the numbers in the question so there will be three in the answer.

4.52 × 3.2
```
        4 5 2
    ×     3 2
  ───────────
        9 0 4
  1 3 5 6 0
  ───────────
  1 4 4 6 4
```

So 4.**52** × 3.2 = 14.4**64**.

Questions

Grade E

1 a Work out each of these.

i 56.2 + 1.6 ii 25.6 + 5.5 iii 32.6 + 3.9

iv 4.9 − 1.3 v 8.43 − 2.6 vi 28.6 − 14.9

b Work out each of these.

i 3.5 × 4 ii 4.7 × 3 iii 3.6 × 3

iv 14.45 ÷ 5 v 15.06 ÷ 6 vi 13.95 ÷ 3

c Work out each of these.

i 4.61 × 23 ii 4.52 × 41 iii 1.72 × 34

iv 3.3 × 0.4 v 0.13 × 0.7 vi 14.2 × 2.8

More fractions

AQA 1/2/3 EDEXCEL 2 OCR 1/2

Adding and subtracting fractions

- When adding and subtracting fractions you must use equivalent fractions to make the denominators of the fractions the same.

$$\frac{3}{8} + \frac{1}{5} = \frac{3 \times 5}{8 \times 5} + \frac{1 \times 8}{5 \times 8} = \frac{15 + 8}{40} = \frac{23}{40}$$

- When adding or subtracting mixed numbers, split the calculation into whole numbers and fractions.

$$4\frac{2}{5} - 1\frac{3}{4} = (4 - 1) + \frac{2}{5} - \frac{3}{4} = 3 + \frac{8}{20} - \frac{15}{20} = 3 + -\frac{7}{20} = 2\frac{13}{20}$$

> **REMEMBER**
> Notice that you may have to split one of the whole numbers if the answer to the fractional part is negative.

Multiplying fractions

- To **multiply** two fractions, multiply the numerators and multiply the denominators.

$$\frac{3}{4} \times \frac{3}{7} = \frac{3 \times 3}{4 \times 7} = \frac{9}{28}$$

- When multiplying **mixed numbers**, change the mixed numbers into **top-heavy** fractions then multiply, as for ordinary fractions.

- **Cancel** any common factors in the top and bottom before multiplying.

- Convert the **final answer** back to a **mixed number** if necessary.

$$3\frac{1}{4} \times 1\frac{1}{5} = \frac{13}{4} \times \frac{6}{5} = \frac{13 \times \overset{3}{\cancel{6}}}{\underset{2}{\cancel{4}} \times 5} = \frac{39}{10} = 3\frac{9}{10}$$

> **REMEMBER**
> On calculator papers you will be expected to use a calculator to do fraction problems, so learn how to use the fraction buttons on your calculator.

Dividing fractions

- To **divide** by a fraction, turn it upside down and multiply by it.

$$\frac{3}{8} \div \frac{7}{9} = \frac{3}{8} \times \frac{9}{7} = \frac{27}{56}$$

- When dividing mixed numbers, change them into top-heavy fractions and then divide, as for ordinary fractions.

- When the second fraction has been turned upside down, cancel before multiplying.

$$1\frac{1}{4} \div 1\frac{7}{8} = \frac{5}{4} \div \frac{15}{8} = \frac{\cancel{5}^1}{\cancel{4}_1} \times \frac{\cancel{8}^2}{\cancel{15}_3} = \frac{2}{3}$$

Questions

Grade C

1 **a** Work out each of these.

    i $\frac{1}{4} + \frac{3}{7}$ **ii** $\frac{5}{6} + \frac{4}{9}$ **iii** $3\frac{2}{3} + 2\frac{2}{5}$

  b Work out each of these.

    i $\frac{3}{5} - \frac{1}{6}$ **ii** $\frac{8}{9} - \frac{2}{3}$ **iii** $2\frac{1}{4} - 1\frac{2}{3}$

  c Work out each of these.

    i $\frac{3}{4} \times \frac{2}{9}$ **ii** $\frac{5}{8} \times \frac{4}{7}$ **iii** $1\frac{2}{5} \times 2\frac{3}{4}$

  d Work out each of these.

    i $\frac{3}{5} \div \frac{6}{7}$ **ii** $\frac{5}{6} \div \frac{10}{21}$ **iii** $3\frac{3}{5} \div 2\frac{1}{4}$

Grade C

PS 2 My son is $\frac{3}{8}$ of my age. My daughter is $\frac{1}{4}$ of my age. Altogether the sum of our ages is 65 years. How old am I?

Number   51

More number

AQA 2   EDEXCEL 2   OCR 2

Multiplying and dividing with negative numbers

- The rules for multiplying and dividing with negative numbers are:
  - when the signs of the numbers are the **same**, the answer is **positive**
  - when the signs of the numbers are **different**, the answer is **negative**.

> **REMEMBER**
> You do not have to write a plus sign in front of a positive number but you must put a minus sign in front of a negative number.

$$+3 \times -4 = -12, \; +12 \div +3 = +4, \; -15 \div -5 = +3, \; -4 \times +6 = -24$$

Rounding to one significant figure

- **Significant figures** are the digits of a number, from the first to the last non-zero digit.

> 3400 has two significant figures (2 sf), 0.06 has one significant figure (1 sf) and 67.45 has four significant figures (4 sf).

- To round a number to one significant figure (1 sf), just round the first two non-zero digits, then replace the rest of the digits in the number with zeros.

> **REMEMBER**
> In GCSE exams you only have to round to one significant figure.

> 432 is 400 to 1 sf, 0.087 is 0.09 to 1 sf and 35.9 is 40 to 1 sf.

Approximation of calculations

- To **approximate** the answer to a calculation, round the numbers in the calculation to one significant figure, then work out the approximate answer.

> 38.2×9.6 can be rounded to 40×10 which is 400, so $38.2 \times 9.6 \approx 400$
>
> $48.3 \div 19.7$ rounds to $50 \div 20 = 2.5$, so $48.3 \div 19.7 \approx 2.5$

- The sign $\approx$ means 'approximately equal to'.

Questions

Grade E

1 a Work out each of these.
  i $+3 \times -5$ **ii** -4×-6 **iii** $-7 \times +5$

  b Work out each of these.
  i $+24 \div -6$ **ii** $-18 \div +9$ **iii** $-12 \div -4$

Grade D

2 a How many significant figures do these numbers have?
  i 6.8   **ii** 0.964   **iii** 120.8

  b Round each of these numbers to one significant figure.
  i 3.8   **ii** 0.752   **iii** 58.7

Grade D

3 Find approximate answers to these calculations.
  a 68.3×12.2
  b $203.7 \div 38.1$
  c $\dfrac{78.3 + 19.6}{21.8 - 9.8}$
  d $\dfrac{42.1 \times 78.6}{4.7 \times 19.3}$

Grade C

4 Find approximate answers to these calculations.
  a $\dfrac{32.1 + 18.4}{0.52}$
  b $\dfrac{98.3 - 29.6}{0.39 - 0.18}$

Number

52

Ratio

Ratio

- A **ratio** is a way of **comparing** the sizes of two or more quantities.

- A **colon** (:) is used to show ratios.   3:4 and 6:20 are ratios.

- Quantities to be compared must be in the **same units** as ratio itself has no units.

- Ratios that have a common factor can be cancelled to give a ratio in its **simplest form**.

- Ratios can be expressed as **fractions**, where the **denominator** is the **sum** of the two parts of the ratio.

> **REMEMBER**
> The method for cancelling a ratio to its simplest form is the same as for cancelling a fraction. Look for the highest common factor.

If a garden has lawn and flower beds in the ratio 3:4 then $\frac{3}{7}$ of the garden is lawn and $\frac{4}{7}$ is flower beds.

Dividing amounts in ratios

- A quantity can be divided into **portions** that are in a certain **given ratio**.

- The process has three steps:

  - **add** the separate parts of the ratio
  - **divide** this number into the original quantity
  - **multiply** this answer by the original parts of the ratio.

> **REMEMBER**
> Always check that the two parts into which you have divided a quantity add up to the original amount.

To share £40 in the ratio 2:3: Add 2 + 3 = 5. Divide 40 ÷ 5 = 8. Multiply each part by 8.
2 × 8 = 16, 3 × 8 = 24. So £40 divided in the ratio 2:3 gives shares of £16 and £24.

Calculating with ratios

- When one part of a ratio is known, it is possible to calculate other values.

- The process has two steps:

  - use the given information to find a **unit value**
  - use the unit value to find the required information.

When the cost of a meal was shared between two families in the ratio 3:5 the smaller share was £22.50. How much did the meal cost altogether?
$\frac{3}{8}$ of the cost was £22.50, so $\frac{1}{8}$ was £7.50. The total cost was 8 × 7.5 = £60.

Questions

Grade D

1 **a** Express each of these as a ratio in its simplest form.

   i 12:36   **ii** 25:30   **iii** 18:30

   b Express each of these as a ratio in its simplest form (remember to express both parts in common units).

   i 50p to £3   **ii** 2 hours to 15 minutes

   iii 40 cm to 1 metre

Grade C

2 Divide the following amounts in the given ratios.

   a £500 in the ratio 1:4

   b 300 grams in the ratio 1:5

   c £400 in the ratio 3:5

   d 240 kg in the ratio 1:2

Grade C

3 **a** A catering box of crisps has two flavours, plain and beef, in the ratio 3:4. There are 42 packets of plain crisps. How many packets of beef crisps are there in the box?

   b The ratio of male teachers to female teachers in a school is 3:7. If there are 21 female teachers, how many teachers are there in total?

Speed and proportion

Speed, time and distance

D

- **Speed**, **time** and **distance** are connected by the formula:

  distance = speed × time

- This formula can be rearranged to give:

  $$\text{speed} = \frac{\text{distance}}{\text{time}} \qquad \text{time} = \frac{\text{distance}}{\text{speed}}$$

- Problems involving speed actually mean **average speed**, as maintaining a constant speed is not possible over a journey.

  > A car travels at 40 mph for 2 hours. How far does it travel in total?
  > Distance = speed × time = 40 × 2 = 80 miles

- Use this diagram to remember the formulae that connect speed, time and distance.

> **REMEMBER**
> Make sure you use the correct units. In speed questions, you are often asked to state the units of your answer.

> **REMEMBER**
> If you are using a calculator, make sure you convert minutes into decimals, for example, 2 hours 15 minutes is 2.25 hours.

Direct proportion problems

D

- When solving **direct proportion** questions, work out the cost of **one item**.

- This is called the **unitary method**.

  > If eight cans of cola cost £3.60, how much do five cans cost?
  >
  > The cost of one can is 360 ÷ 8 = 45p, so five cans cost 5 × 45p = £2.25.

> **REMEMBER**
> Always check that the answer is sensible and makes sense, compared with the numbers in the original problem.

Best buys

D

- Many products are sold in **different sizes** at different **prices**.

- To find a **best buy**, work out how much of the item you get for **a unit** cost, such as how much you get **per penny** or **per pound**.

- Always **divide** the **quantity** by the **cost**.

  > A 400 g jar of coffee costs £1.44. A kilogram jar of the same coffee costs £3.80. Which jar is the better value?
  > 400 ÷ 144 = 2.77 g/penny   1000 ÷ 380 = 2.63 g/penny
  > Hence, the 400 g jar is better value.

> **REMEMBER**
> Be careful with units. Change pounds into pence and kilograms into grams.

Questions

Grade D

FM 1 a A motorist travels a distance of 75 miles in 2 hours. What is the average speed?

  b A cyclist travels for $3\frac{1}{2}$ hours at an average speed of 15 km per hour. How far has she travelled?

Grade C

2 a 40 bricks weigh 50 kg. How much will 25 bricks weigh?

  b How many bricks are there on a pallet weighing 200 kg, if you ignore the weight of the pallet?

Grade C

FM 3 a A large tube of toothpaste contains 250 g and costs £1.80. A travel-size tube contains 75 g and costs 52p. Which is the better value?

  b Which is the better mark:

  62 out of 80

  95 out of 120?

Percentages

Equivalent percentages, fractions and decimals

- **Fractions**, **percentages** and **decimals** are all different ways of expressing parts of a whole.

- This table shows how to convert from one to another.

	Decimal	Fraction	Percentage
Percentage to:	Divide by 100 (or move the decimal point two places left). $60\% = 0.6$	Put it over 100 and cancel if possible. $55\% = \frac{55}{100} = \frac{11}{20}$	
Fraction to:	Divide the numerator by the denominator. $\frac{4}{5} = 4 \div 5 = 0.8$		Divide the numerator by the denominator and multiply by 100. $\frac{7}{8} = 7 \div 8 \times 100$ $= 87.5\%$
Decimal to:		Make the denominator 10 or 100 then cancel if possible. $0.68 = \frac{68}{100} = \frac{17}{25}$	Multiply by 100 (or move the decimal point two places right). $0.76 = 76\%$

> **REMEMBER**
> Most of the questions in GCSE use very basic values so it is worth learning some of them, such as
> $0.1 = \frac{1}{10} = 10\%$,
> $0.25 = \frac{1}{4} = 25\%$.

The percentage multiplier

- Using the **percentage multiplier** is the best way to solve percentage problems.

- The percentage multiplier is the **percentage** expressed as a **decimal**.

  72% gives a multiplier of 0.72, 20% is a multiplier of 0.20 or 0.2.

- The multiplier for a percentage **increase** or **decrease** is the percentage multiplier **added to 1** or **subtracted from 1**.

  An 8% increase is a multiplier of 1.08 (1 + 0.08), a 5% decrease is a multiplier of 0.95 (1 − 0.05).

> **REMEMBER**
> Learn how to use multipliers as they make percentage calculations easier and more accurate.

Questions

Grade D

1 a Write the following percentages as decimals.
   i 30%   ii 88%

  b Write the following percentages as fractions.
   i 90%   ii 32%

  c Write the following decimals as percentages.
   i 0.85   ii 0.15

  d Write the following decimals as fractions.
   i 0.8   ii 0.08

  e Write the following fractions as decimals.
   i $\frac{5}{8}$ ii $\frac{7}{20}$

  f Write the following fractions as percentages.
   i $\frac{9}{25}$ ii $\frac{1}{20}$

Grade D

2 a Write down the percentage multiplier for each of these.
   i 80%   ii 7%   iii 22%

  b Write down the multiplier for each percentage increase.
   i 5%   ii 12%   iii 3.2%

  c Write down the multiplier for each percentage decrease.
   i 8%   ii 15%   iii 4%

Number

Calculating a percentage of a quantity

- To calculate a percentage of a quantity simply **multiply** the quantity by the **percentage multiplier**.

  What is 8% of 56 kg?

  Work it out as 0.08 × 56 = 4.48 kg.

REMEMBER

On the non-calculator paper, once you have 10%, you can find 5% or 20% easily by dividing or multiplying by 2.

- On the non-calculator paper, percentages will be **based on 10%**.
- To work out 10%, just **divide the quantity by 10**.

Percentage increase or decrease

- To calculate the new value after a quantity is **increased** or **decreased** by a percentage, simply **multiply** the original quantity by the **percentage multiplier** for the increase or decrease.

  What is the new cost after a price of £56 is decreased by 15%?
  Work it out as 0.85 × 56 = £47.60.

  Jamil gets a wage increase of 4%. His previous wage was £180 per week. How much does he get now?
  Work it out as 180 × 1.04 = £187.20.

REMEMBER

If the calculator shows a number such as 47.6 as the answer to a money problem, always put the extra zero into the answer, so you write this down as £47.60.

One quantity as a percentage of another

- To calculate one **quantity** as a **percentage** of **another** just divide the first quantity by the second. This will give a decimal, which can be converted to a percentage.

  A plant grows from 30 cm to 39 cm in a week. What is the percentage growth?

  The increase is 9 cm, 9 ÷ 30 = 0.3 and this is 30%.

REMEMBER

Always divide by the original quantity, otherwise you will get no marks.

Questions

Grade E

1 a Work these out.
   i   15% of £70   ii   32% of 60 kg
  b A jacket is priced at £60. Its price is reduced by 12% in a sale.
   i   What is 12% of £60?
   ii What is the sale price of the jacket?

Grade D

2 a Increase £150 by 12%.
  b Decrease 72 kg by 8%.

Grade C

3 a The average attendance at Barnsley football club in 2005 was 14 800.

   In 2006 it was 15 540.

   i   By how much had the average attendance gone up?
   ii What is the percentage increase in attendance?

  b After her diet, Carol's weight had gone from 80 kg to 64 kg.

   What is the percentage decrease in her weight?

Grade C

AU 4 Explain why a 5% increase followed by a 6% decrease is the same as a 6% decrease followed by a 5% increase.

Number grade booster

I can...

- [ ] recall the times tables up to 10×10
- [ ] use BODMAS to do calculations in the correct order
- [ ] identify the place value of digits in whole numbers
- [ ] round numbers to the nearest 10 or 100
- [ ] add and subtract numbers with up to four digits without a calculator
- [ ] multiply numbers by a single-digit number
- [ ] state what fraction of a shape is shaded
- [ ] shade in a fraction of a shape
- [ ] add and subtract simple fractions with the same denominator
- [ ] recognise equivalent fractions
- [ ] cancel a fraction
- [ ] change top-heavy fractions into mixed numbers and vice versa
- [ ] find a fraction of an integer
- [ ] recognise the multiples of the first 10 whole numbers
- [ ] recognise square numbers up to 100
- [ ] find equivalent fractions, decimals and percentages
- [ ] recognise that a number on the left on a number line is smaller than a number on the right
- [ ] read negative numbers on scales such as thermometers

You are working at **Grade G** level.

- [ ] divide numbers by a single-digit number
- [ ] put fractions in order of size
- [ ] add fractions with different denominators
- [ ] solve fraction problems expressed in words
- [ ] compare fractions of quantities
- [ ] find factors of numbers less than 100
- [ ] add and subtract with negative numbers
- [ ] write down the squares of numbers up to 15×15
- [ ] write down the cubes of 1, 2, 3, 4, 5 and 10
- [ ] use a calculator to find square roots
- [ ] do long multiplication
- [ ] do long division
- [ ] solve real-life problems involving multiplication and division
- [ ] round decimal numbers to one, two or three decimal places
- [ ] find percentages of a quantity
- [ ] change mixed numbers into top-heavy fractions

You are working at **Grade F** level.

☐ multiply fractions

☐ add and subtract mixed numbers

☐ calculate powers of numbers

☐ recognise prime numbers under 100

☐ use the four rules with decimals

☐ change decimals to fractions

☐ change fractions to decimals

☐ simplify a ratio

☐ find a percentage of any quantity

You are working at **Grade E** level.

☐ work out one quantity as a fraction of another

☐ solve problems using negative numbers

☐ multiply and divide by powers of 10

☐ multiply together numbers that are multiples of powers of 10

☐ round numbers to one significant figure

☐ estimate the answer to a calculation

☐ order lists of numbers containing decimals, fractions and percentages

☐ multiply and divide fractions

☐ calculate with speed, distance and time

☐ compare prices to find 'best buys'

☐ find the new value after a percentage increase or decrease

☐ find one quantity as a percentage of another

☐ solve problems involving simple negative numbers

☐ multiply and divide fractions

You are working at **Grade D** level.

☐ work out a reciprocal

☐ recognise and work out terminating and recurring decimals

☐ write a number as a product of prime factors

☐ use the index laws to simplify calculations and expressions

☐ multiply and divide with negative numbers

☐ multiply and divide with mixed numbers

☐ find a percentage increase

☐ work out the LCM and HCF of two numbers

☐ solve problems using ratio in appropriate situations

You are working at **Grade C** level.

Perimeter and area

Perimeter

- The **perimeter** of a shape is the **sum** of the lengths of all its **sides**.

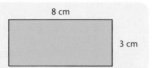

The perimeter of this rectangle is 22 cm.

8 cm

3 cm

- A **compound shape** is a 2-D shape that is made up of other simple shapes, such as rectangles and triangles.

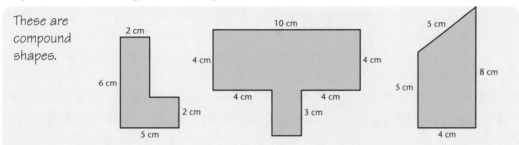

These are compound shapes.

2 cm

6 cm

2 cm

5 cm

10 cm

4 cm

4 cm   4 cm

3 cm

4 cm

5 cm

5 cm

5 cm

8 cm

4 cm

Area of irregular shapes (counting squares)

- The basic **units** of area are square centimetres (**cm²**), square metres (**m²**) and square kilometres (**km²**).

- To work out the area of an irregular shape, trace it onto a sheet of **centimetre squares**, say, and then **count** the squares.

These shapes are drawn on a centimetre grid.

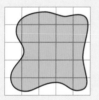 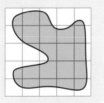

Area of a rectangle

- The area of a rectangle can be calculated by the formula:

area = length × width $A = lw$

The area of the rectangle above is:
$A = 8 \times 3 = 24 \text{ cm}^2$

Questions

Grade G

1   Find the perimeter of each of the compound shapes shown above.

Grade G

2   By counting squares, find the area of the two irregular shapes above.

Grade F

3   a   Find the area of a rectangle with sides of 6 m and 2.5 m.

    b   Find the area of a rectangle with sides of 3.2 cm and 5 cm.

Geometry

Area of a compound shape

- To find the area of a compound shape, **divide** the shape into regular shapes, such as rectangles and triangles, and calculate the areas of the separate parts.

> **REMEMBER**
> The sides of the rectangles into which you split the compound shape will probably need to be calculated by subtraction.

Find the area of this shape.

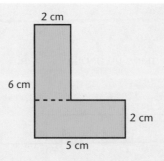

The area of the top rectangle is $4 \times 2 = 8 \text{ cm}^2$

The area of the bottom rectangle is $5 \times 2 = 10 \text{ cm}^2$

The total area is $8 + 10 + 18 \text{ cm}^2$

Area of a triangle

- To work out the area of a triangle use the formula:

area $= \frac{1}{2} \times$ **base** $\times$ **height**

$A = \frac{1}{2} bh$

Find the area of this triangle.

> **REMEMBER**
> Sometimes in exams one of the sides of the triangle is included even though it is not needed as part of the calculation. This is to test if you can pick out the correct value for the height.

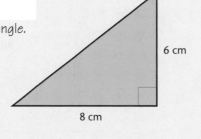

$A = \frac{1}{2} \times 8 \times 6 = 4 \times 6 = 24 \text{ cm}^2$

- The height of a triangle is the **perpendicular distance** from the base to the top point or **vertex**.

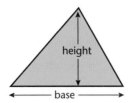

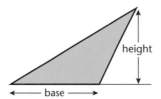

Questions

1 Find the area of a triangle with:
   a base 7 cm, height 10 cm
   b base 5 m, height 5 m.

2 Find the area of each of these two compound shapes.

a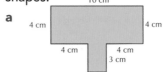

b

Geometry

Area

Area of a parallelogram

- To find the **area** of a parallelogram, **multiply** the **base** by the **perpendicular distance** between the sides (usually called the **height**).

- The formula for the area of a parallelogram is usually given as:

area = base × height

$A = bh$

Find the area of this parallelogram.

3 cm

8 cm

Area = 8 × 3 = 24 cm²

> **REMEMBER**
> Very often the slant side of a parallelogram is included even though it is not needed to calculate the area. This is to test if you can ignore unnecessary information.

Area of a trapezium

- To work out the area of a trapezium, multiply half the sum of the parallel sides by the distance between them.

- Use the formula:

area, $A = \frac{1}{2} \times h \times (a + b)$

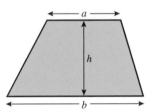

> **REMEMBER**
> The formula for the area of a trapezium is given on the formula sheet that is included with the exam. However, it is much better to learn it.

Find the area of this trapezium.

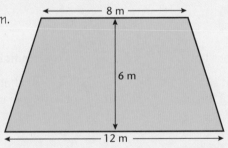

8 m

6 m

12 m

$A = \frac{1}{2} \times (8 + 12) \times 6 = \frac{1}{2} \times 20 \times 6 = 60$ m²

Questions

1 Find the area of each of these parallelograms.

a

2 cm

10 cm

b

5 m

7 m

2 Find the area of each of these trapezia.

a

5 cm

5 cm

10 cm

b

7 cm

5 cm

4 cm

10 cm

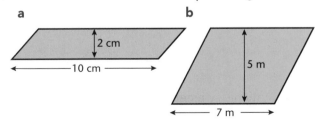

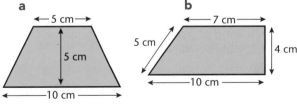

Symmetry

F

Lines of symmetry

- Many two-dimensional shapes have one or more **lines of symmetry**.

- A **line of symmetry** is a line that can be drawn through a shape so that what can be seen on one side of the line is the **mirror image** of what is on the other side.

- Lines of symmetry are also called **mirror lines**.

REMEMBER

The easiest way to check a shape for a line of symmetry is to trace it and fold the tracing paper along the mirror line. The object and its image will be on top of each other.

These shapes all have lines of symmetry.

F

Rotational symmetry

- A two-dimensional shape has **rotational symmetry** if it can be **rotated** about a **point** to look exactly the same in a new position.

- The **order** of rotational symmetry is the **number of different positions** in which the shape looks the **same** when it is rotated.

REMEMBER

The easiest way to check for rotational symmetry is to trace the shape and, as you rotate the tracing paper a full turn about the centre, count how many times it looks the same.

These shapes all have rotational symmetry.

D

Planes of symmetry

- Many two-dimensional shapes have one or more **planes of symmetry**.

- A **plane of symmetry** is a flat surface that can be cut through a shape so that what is on one side of the plane is the **mirror image** of what is on the other side.

REMEMBER

Imagine the solids are made of modelling clay and you are cutting them with a knife.

These solids all have planes of symmetry.

Cube   Triangular prism   Square-based pyramid

Questions

Grade D

1 Use tracing paper or otherwise to find out how many lines of symmetry the plane shapes, above (top), have.

Grade D

2 Use tracing paper or otherwise to find out the order of rotational symmetry the plane shapes, above (middle), have.

Grade D

3 How many planes of symmetry do the solids shown above have?

Geometry

Angles

Measuring and drawing angles

- To measure an angle, use a **protractor**.

- A **half round** protractor measures up to 180° and a **full round** protractor measures up to 360°.

- It is important to remember two rules when measuring angles:

  - put the **centre of the protractor** exactly over the **centre (vertex) of the angle**

  - make sure the **zero line** of the protractor is exactly **along one of the arms** of the angle.

> **REMEMBER**
> Full round protractors are best, especially for working with bearings, which come later.

To measure reflex angles, such as GHI shown below, it is easier to use a circular protractor.

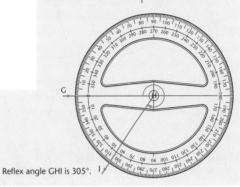

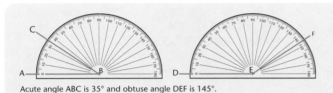

Acute angle ABC is 35° and obtuse angle DEF is 145°.

Reflex angle GHI is 305°.

Angle facts

- There are many angle facts that can be used to solve problems:

  - **angles on a straight line** add up to 180°

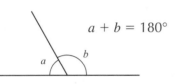

$a + b = 180°$

$c + d + e + f = 180°$

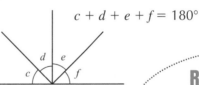

  - **angles around a point** add up to 360°.

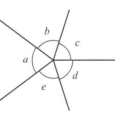

$a + b + c + d + e = 360°$

> **REMEMBER**
> If you are asked to give reasons for how you worked out an angle, write 'angles on a straight line' or 'angles around a point' or 'angles in a full turn'. You do not need to write any more than that.

Questions

Grade F

1 Measure these angles.

a

b

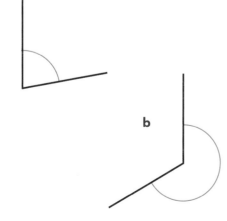

Grade F

2 What is the value of each of the angles marked with letters?

a

a 52°

b b 81°

73°

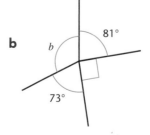

Geometry

Angles in a triangle

- The **angles in a triangle** add up to **180°**.
- There are four types of triangle.

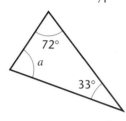

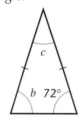

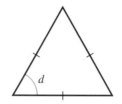

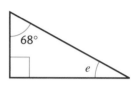

A **scalene** triangle doesn't have any two angles the same or any two sides the same.

An **isosceles** triangle has two angles the same and two sides the same.

An **equilateral triangle** has all three angles the same (60°) and all three sides the same length.

A **right-angled** triangle has one right angle (90°).

Angles in a quadrilateral

- The angles in a **quadrilateral** add up to **360°**.
- Any quadrilateral can be formed from **two triangles**.

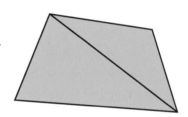

Some quadrilaterals are shown below.

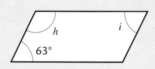

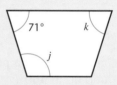

Square   Rectangle   Parallelogram   Isosceles trapezium

Questions

Grade E

1 Find the values of the angles marked with letters in the triangles, above.

Grade D

AU 2 Use the diagram, above, of the quadrilateral with one diagonal to explain why the angles in a quadrilateral add up to 360°.

Grade D

3 Find the values of the angles marked with letters in the quadrilaterals, above.

Grade D

4 Write down the value of the angles a, b, c, d and e in the kite and the rhombus shown.

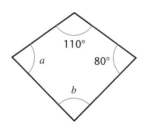

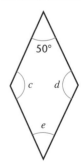

Polygons

Polygons

- The angles in a **triangle** add up to **180°**.
- The angles in a **quadrilateral** add up to **360°**.
- The angles in a **pentagon** add up to **540°**.
- The angles in a **hexagon** add up to **720°**.
- The angle sum increases by 180° for each side added because each time an extra triangle is added.
- In **any polygon** there are **two fewer triangles** than the **number of sides**.

REMEMBER

If you are asked to find the angles in a polygon, split it into triangles.

Triangle

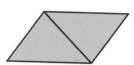

Quadrilateral

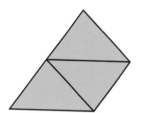

Pentagon

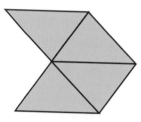

Hexagon

Regular polygons

- A **regular polygon** is a polygon with all its sides the same length.
- In any regular polygon, all the **interior angles**, i, are equal, and all of the **exterior angles**, e, are equal.

Here are three regular polygons.

Interior and exterior angles in a regular polygon

- The **interior** and **exterior** angles of some regular polygons are shown below.
- To find the interior angle of any regular polygon, **divide the angle sum** by the number of **sides**.
- To find the exterior angle of any regular polygon, **divide 360°** by the number of **sides**.

Equilateral triangle

Square

Pentagon

Hexagon

Octagon

Questions

Grade C

1 What is the interior angle of a regular: **a** octagon   **b** nonagon?

Grade C

2 What is the exterior angle of a regular: **a** octagon   **b** nonagon?

Grade C

AU 3 What is the connection between the interior and exterior angles of any regular polygon?

Parallel lines and angles

AQA 3   EDEXCEL 2   OCR 2

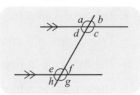

D

Two parallel lines and a transversal

- A line that **crosses** two **parallel lines** it is called a **transversal**.
- At each **crossing point** there are **four angles**, so there are eight angles in total.
- These angles have some **mathematical relationships** and **special names**.

C

Alternate angles

- **Alternate angles** are angles that are on opposite sides of the transversal within the parallel lines.
- Alternate angles are **equal**.

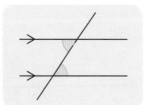

> **REMEMBER**
> Alternate angles are sometimes called Z angles, as they form a 'Z' shape, but *do not* use 'Z angles' as a reason in an exam or you will not get full marks.

C

Corresponding angles

- **Corresponding angles** are angles that are on the same side of the transversal and the parallel lines.
- Corresponding angles are **equal**.

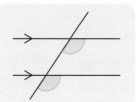

> **REMEMBER**
> Corresponding angles are sometimes called F angles, as they form an 'F' shape, but *do not* use 'F angles' as a reason in an exam or you will not get full marks.

D

Opposite angles

- **Opposite angles** (also known as **vertically opposite**) are angles that are opposite each other across the intersection of two lines.
- Opposite angles are **equal**.

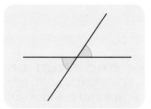

D

Interior angles

- **Interior angles** (also known as **co-interior** or **allied angles**) are angles that are on the same side of the transversal and within the parallel lines.
- Interior angles **add up to 180°**.

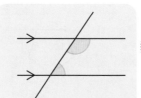

> **REMEMBER**
> Interior angles are sometimes called C or U angles as they form a 'C' or 'U' shape but *do not use* 'C angles' as a reason in an exam or you will not get full marks.

Questions

Grade C

1 Refer to the top diagram, above.
  a Which angle is alternate to angle f?
  b Which angle is alternate to angle c?
  c Which angle is corresponding to angle f?
  d Which angle is corresponding to angle a?
  e Which angle is interior to angle f?
  f Which angle is opposite to angle f?
  g Which angle is opposite to angle a?
  h Which angle is interior to angle d?

Quadrilaterals

Special quadrilaterals

- There are many different quadrilaterals. You will already know the **square** and the **rectangle**.

- The **square** has the following properties:
  - all sides are equal
  - all angles are equal
  - diagonals bisect each other and cross at right angles.

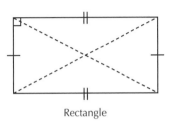

Square

> **REMEMBER**
> If you are asked about a quadrilateral, draw it and draw in the diagonals. Then mark on all the equal angles and sides.

- The **rectangle** has the following properties:
  - opposite sides are equal
  - all angles are equal
  - diagonals bisect each other.

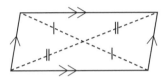

Rectangle

- The **parallelogram** has the following properties:
  - opposite sides are equal and parallel
  - opposite angles are equal
  - diagonals bisect each other.

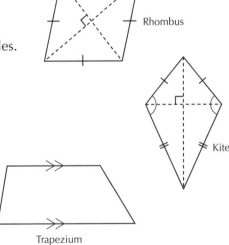

Parallelogram

- The **rhombus** has the following properties:
  - all sides are equal
  - opposite sides are parallel
  - opposite angles are equal
  - diagonals bisect each other and cross at right angles.

Rhombus

- The **kite** has the following properties:
  - two pairs of sides are equal
  - one pair of opposite angles are equal
  - diagonals cross at right angles.

Kite

- The **trapezium** has the following properties:
  - two unequal sides are parallel
  - interior angles add to 180°.

- An **isosceles trapezium** has a **line of symmetry**.

Trapezium

Questions

Grade C

1 Which quadrilaterals have the following properties?
  a Diagonals cross at right angles.
  b Diagonals bisect each other.
  c All sides are equal.
  d Both pairs of opposite sides are equal.
  e Both pairs of opposite angles are equal.
  f One pair of sides are parallel.
  g One pair of angles are equal.

Grade C

AU 2 a Marcie says, 'All squares are rectangles.' Is she correct?
  b Milly says, 'All rhombuses are parallelograms.' Is she correct?
  c Molly says, 'All kites are rhombuses.' Is she correct?
  d Tilly says 'All squares are rhombuses'. Is she correct?

Bearings

Bearings

- **Bearings** are used to describe the position and direction of one object in relation to another object.

- Bearings are used to describe positions of aircraft or ships, for example. Hikers also use bearings to make sure they do not get lost in bad weather.

- The **bearing** of an object, from where you are standing, is the angle through which you turn towards the object, in a clockwise direction, as you turn from facing north.

- Bearings are also known as **three-figure bearings** as it is normal to give a bearing of 60° as 060°.

- The main points of the **compass rose** have bearings of 000°, 045°, 090°, etc.

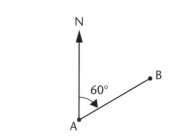

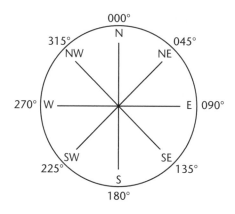

Measuring a bearing

- Draw a north line.

  – Place the centre of the protractor exactly on the point from which you are measuring.

  – Line up the 0° line with north.

  – Read the bearing, using the clockwise scale.

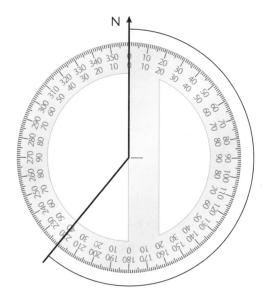

Questions

Grade D

1 What bearing is being measured by the protractor, above?

Grade D

2 a What is the **opposite** bearing to east?

  b What is the **opposite** bearing to north-east?

Grade D

3 A plane is flying on a bearing of 315°. In what compass direction is it flying?

Grade C

AU 4 On a flat desert surface a man walks 1 kilometre north, then 1 kilometre west.

He then heads directly back to where he started.

On what bearing will he be walking?

Circles

Circles

- You need to know the terms, labelled on the diagram, that relate to **circles**.

- You also need to know about π (pronounced '**pi**'), a special number used in work with circles.

- The value of p is a **decimal** that goes on for ever but it is taken as 3.142 to three decimal places.

- All calculators should have a π **button** that will give the value as 3.141 592 654.

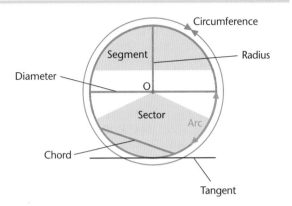

Circumference of a circle

- The **circumference** of a circle is the distance around the circle (the **perimeter**).

- The circumference of a circle is given by the formula:

$C = \pi d$ or $C = 2\pi r$

where d is the **diameter** and r is the **radius**.

- The formula you use depends on whether you are given the radius or the diameter.

> What is the circumference of a circle with a radius of 20 cm?
> $C = 2 \times \pi \times 20 = 40\pi = 125.7$ cm

- Sometimes on the non-calculator paper you are asked to give an answer in terms of π. In this case leave your answer as, for example, 40π.

> **REMEMBER**
> You can learn just one formula because you can always find the radius from the diameter, or vice versa, as: $d = 2r$

> **REMEMBER**
> Unless you are asked to give your answer in terms of π, use a calculator to work out circumferences and areas of circles and give answers correct to at least one decimal place.

Area of a circle

- The area of a circle is the **space inside** the circle.

- The area of a circle is given by the formula:

$A = \pi r^2$

where r is the **radius**.

- You must use the radius when calculating the area.

Questions

Grade D

1 a Calculate the circumference of a circle with radius 7 cm. Give your answer to one decimal place.

   b Calculate the circumference of a circle with diameter 12 cm. Give your answer to one decimal place.

   c Calculate the circumference of a circle with radius 4 cm. Leave your answer in terms of π.

Grade D

2 a Calculate the area of the circle, above, with diameter 8 cm.
      Give your answer to one decimal place.

   b Calculate the area of a circle with radius 15 cm.
      Give your answer to one decimal place.

   c Calculate the area of a circle with radius 3 cm.
      Leave your answer in terms of π.

Scales

G

- You will come across **scales** in many places in everyday life, for example, on thermometers, car speedometers and on kitchen scales.

- When you read a scale, make sure you know what each **division** on the scale represents.

- When you read a scale, make sure that you read it in the **right direction**.

> **REMEMBER**
> Questions that ask you to read a scale often ask you to include the units in your answer.

Sensible estimates

F

- You should know some **basic measurements**, for example, the **average height** of a man is about 1.8 metres and **doorways** are about 2 metres high and **a bag of sugar** weighs 1 kilogram.

- You can use this **basic information** to make **estimates** of other lengths, heights and weights.

> **REMEMBER**
> As you are being asked for an estimate there will be a range of acceptable values.

Questions

Grade G

1 Include the units of your answer.

  a What temperature is being shown on the thermometer, above?

  b What speed is being shown on the car speedometer, above?

  c What weight is being shown on the kitchen scales, above?

Grade G

2 a From the picture, above, estimate:

    i the height of the lamppost

    ii the length of the bus.

  b In the scales in the picture, above, three textbooks are balanced by four kilogram-bags of sugar. Estimate the weight of one textbook.

Scales and drawing

Scale drawing

- Scale drawings are used to give an **accurate representation** of a real object.

> This scale drawing shows the plan of a bedroom.
>
>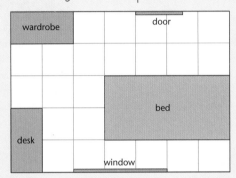
>
> Scale: 1 cm represents 50 cm

> **REMEMBER**
> Be careful with units. All units must be the same so convert metres to centimetres, for example.

- When you use a scale drawing make sure you know what the scale is.

- The actual measurement and the scaled measurement are connected by a **scale factor**.

- Scales are often given as **ratios**.

> The wardrobe is 2 centimetres by 1 centimetre on the scale drawing.
> The scale is 1 centimetre represents 50 centimetres, or 1 : 50.
> So the wardrobe is 1 metre by 50 centimetres in the real bedroom.

Nets

- **Polyhedra** are solid shapes with **plane** (flat) sides.

- A **net** is a flat shape that can be folded into the **3-D shape**.

- Most **three-dimensional shapes**, particularly polyhedra, can be made from **nets**.

> These are the nets of two common solids.
>
>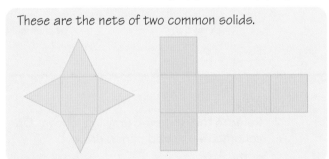

Questions

Grade F

1 Refer to the scale drawing of the bedroom, above.

   a How wide is the window?

   b What are the dimensions of the bed?

   c The desk is to be covered with a plastic sheet. The plastic costs £5 per square metre. How much will it cost to cover the desk?

Grade F

2 What two solids do the nets, above, represent?

Grade E

AU 3 Which of these nets make a cube?

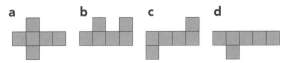

a   b   c   d

Isometric grids

- **Isometric grids** are usually drawn with solid lines or as a **dotty triangular grid**.

- An isometric grid can be used to show a **two-dimensional view** of a **three-dimensional object**.

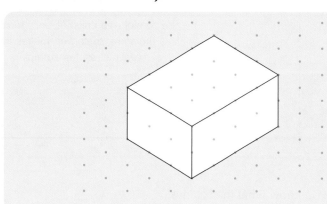

REMEMBER

When questions using isometric grids are set in examinations the grid will be the right way round. Lines are always drawn in three directions only.

- When you use isometric grids it is important to ensure they are the **correct way round**, as in the diagram above.

Plans and elevations

- The **plan** of a shape is the view seen when looking directly down from above.

- The **front elevation** of a shape is the view seen when looking from the front.

- The **side elevation** of a shape is the view seen when looking from the side.

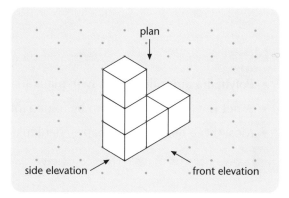

Questions

Grade E

1 Refer to the isometric drawing of the cuboid, above.

  a What are the lengths of the sides of the cuboid?

  b What is the volume of the cuboid?

Grade D

2 Refer to the isometric drawing, above. On centimetre-squared paper, draw:

  a the plan

  b the front elevation

  c the side elevation.

Congruency and tessellations

Congruent shapes

- When two shapes are **congruent** they have **exactly the same** dimensions.
- Congruent shapes can be **reflections** or **rotations** of each other.

These triangles are all congruent to each other.

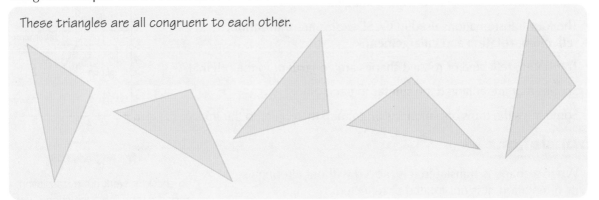

Tessellations

- When a shape **tessellates** it fits together so that there are **no overlaps** and **no gaps**.
- A **tessellation** is the pattern that is formed when the shapes are fitted together.

These are all tessellations.

REMEMBER

To check congruency, use tracing paper to copy one shape and place the tracing paper over the other shapes to see if they are exactly the same. You may have to turn the tracing paper over.

Questions

Grade G

1 Are the shapes in each pair congruent?

a

b

c

Grade E

2 The diagram shows a tessellation of a 1 cm by 2 cm rectangle.

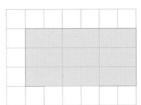

On centimetre-squared paper, draw a different tessellation of a 1 cm by 2 cm rectangle. Draw at least six rectangles to show your tessellation clearly.

Transformations

Transformations

D

- When a shape is **moved**, **rotated**, **reflected** or **enlarged**, this is a **transformation**. The original shape is called the **object**, and the transformed shape is called the **image**.

- The four transfomations used in GCSE exams are **translation**, **reflection**, **rotation** and **enlargement**.

- Translated, reflected or rotated shapes are **congruent** to the original.

- Shapes that are enlarged are **similar** to each other.

- Sometimes the transformation is said to **map** the object to the image.

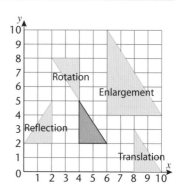

Translations

D

- When a shape is **translated** it is moved without altering its orientation; it is not rotated or reflected.

- A translation is described by a **vector**, for example, $\binom{-4}{5}$.

- The **top number** in the vector is the movement in the x-direction. Positive values move to the right. Negative values move to the left.

- The **second** or **bottom number** in the vector is the movement in the y-direction, or vertically. Positive values move upwards. Negative values move downwards.

> **REMEMBER**
> To check or work out a translation, use tracing paper. Trace the shape and then count squares as you move it horizontally and vertically.

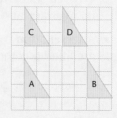

> Triangle C is translated from triangle A by the vector $\binom{0}{4}$.
> Triangle B is translated to triangle D by the vector $\binom{-2}{4}$.

Reflections

D

- When a shape is **reflected** it becomes a **mirror image** of itself.

- A **reflection** is described in terms of a mirror line.

- **Equivalent points** on either side are the **same distance** from the mirror line and the **line joining them** crosses the **mirror line** at **right angles**.

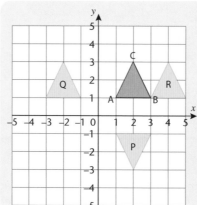

> **REMEMBER**
> The mirror lines in GCSE questions will always be of the form $y = a$, $x = b$, $y = x$, $y = -x$.

> Triangle P is a reflection of the shaded triangle in the x-axis.

Questions

Grade D

1 Refer to the middle diagram, above.
What vector translates:
  a triangle A to **i** triangle B **ii** triangle D
  b triangle C to **i** triangle B **ii** triangle A?

Grade D

2 Refer to the bottom diagram, above.
What is the mirror line for the reflection:
  a that takes the shaded triangle to triangle Q
  b that takes the shaded triangle to triangle R?

Geometry

Rotations

- When a shape is **rotated** it is turned about a centre, called the **centre of rotation**.

- The rotation will be in a **clockwise** or **anticlockwise** direction.

- The rotation can be described by an **angle**, such as 90°, or a **fraction of a turn**, such as 'a half-turn'.

> **REMEMBER**
> To check or work out a rotation, use tracing paper. Use a pencil point on the centre and rotate the tracing paper in the appropriate direction, through the given angle.

Triangle A is a rotation of the shaded triangle through 90° in a clockwise direction about the centre (0, 1).

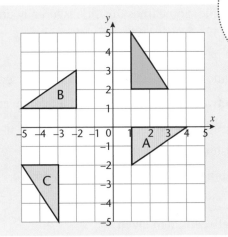

- A rotation of 90° clockwise is the same as a rotation of 270° anticlockwise. Only a half-turn does not need to have a direction specified.

Enlargements

- When a shape is **enlarged** it changes its size to become a shape that is **similar** to the first shape.

> **REMEMBER**
> Scale factors can also be fractions. In this case the enlargement makes things smaller!

- An **enlargement** is described by a **centre** and a **scale factor**.

- The length of the sides of the image will be the length of the sides of the object **multiplied** by the **scale factor**.

- The centre of enlargement can be found by the **ray method**.

- The ray method involves drawing lines through corresponding points in the object and image. The point where the **rays meet** is the **centre of enlargement**.

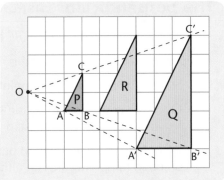

Triangle Q is an enlargement of triangle P with a scale factor of 3 from the centre O.

Questions

Grade D

1 Refer to the top diagram, above. What is the rotation that takes:

   a the shaded triangle to triangle B

   b the shaded triangle to triangle C?

Grade D

2 Refer to the bottom diagram, above.

   a i Write down the scale factor of the enlargement that takes triangle P to triangle R.

   ii Mark the centre of the enlargement that takes triangle P to triangle R.

   b i Write down the scale factor of the enlargement that takes triangle Q to triangle R.

   ii Mark the centre of the enlargement that takes triangle Q to triangle R.

   iii State the scale factor and centre of the enlargement that takes triangle Q to triangle P.

Constructions

D

Constructing triangles

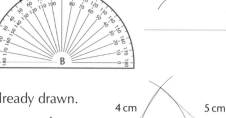

- When you are asked to **construct** a triangle you are expected to use a pair of **compasses** to measure lengths and a **protractor** to measure angles.

- There are three ways of constructing triangles.

- **All three sides given**

  - Use a ruler to draw one side. Sometimes this side is already drawn.

  - Use compasses to measure the other two sides and draw arcs from the ends of the side you have drawn.

  - Join up the ends of the line to the point where the arcs intersect.

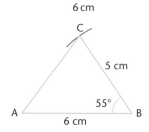

- **Two sides and the angle between them (the included angle) given**

  - Use a ruler to draw one side. Sometimes this side is already drawn.

  - Use a protractor to measure and draw the angle.

  - Use compasses to measure the other length and draw an arc.

  - Join up the points.

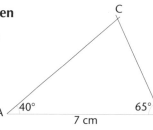

- **Two angles and the side between them given**

  - Use a ruler to draw the side. Sometimes this side is already drawn.

  - Use a protractor to measure and draw the angles at each end of the line.

  - Extend the lines to form the triangle.

> **REMEMBER**
> Always show your construction lines and arcs clearly. You will not get any marks if they can't be seen.

The perpendicular bisector

C

- To **bisect** means to *divide in half*. **Perpendicular** means *at right angles*.

- A **perpendicular bisector** divides a line in two, and is at right angles to it.

  - Start off with a line or two points. (Usually, these are given.)

  - Open the compasses to a radius about three-quarters of the length of the line, or the distance between the points.

  - With the compasses centred on each end of the line (or each point) in turn, draw arcs on either side of the line.

  - Join up the points where the arcs cross.

Questions

Grade D

1 Use a ruler, compasses and protractor for these questions.

   a Draw accurately the triangle with sides of 4 cm, 5 cm and 6 cm, shown above.

   b Draw accurately the triangle with sides of 6 cm and 5 cm and an included angle (between them) of 55°, shown above.

   c Draw accurately the triangle with a side of 7 cm, and angles of 40° and 65°, shown above.

Grade C

2 Draw a line 6 cm long. Following the steps above, draw the perpendicular bisector of the line. Check that each side is 3 cm long and that the angle is 90°.

Construction and loci

AQA 3   EDEXCEL 3   OCR 1

The angle bisector

- An **angle bisector** divides an angle into two smaller, equal angles.

  - Start off with an angle. (Usually, this is given.)

  - Open the compasses to any radius shorter than the arms of the angle. With the compasses centred on the **vertex** of the angle (the point where the arms meet), draw an arc on each arm of the angle.

  - Now, with the compasses still set to the same radius, draw arcs from these arcs so they intersect.

  - Join up the vertex of the angle and the point where the arcs cross.

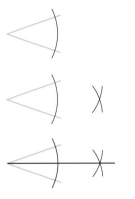

Loci

- A **locus** (plural **loci**) is the path followed by a point according to a rule.

  A point that moves so that it is always 4 cm from a fixed point is a circle of radius 4 cm.

  A point that moves so that it is always the same distance from two fixed points is the perpendicular bisector of the two points.

Practical problems

- **Loci** can be used to solve real-life problems.

  A horse is tethered to a rope 10 m long in a large, flat field.
  What is the area the horse can graze?

  The horse will be able to graze anywhere within a circle of radius 10 m.

REMEMBER

These questions are often not answered well in examinations because candidates forget the techniques. The night before the exam, look up 'Euclidean constructions' on the internet to find websites that will give you a dynamic demonstration of all the constructions you need for GCSE.

- In reality the horse may not be able to graze an exact circle but the situation is **modelled** by the mathematics.

Questions

Grade C

1 Draw an angle of 70°. Following the steps above, draw the angle bisector of the line. Check that each half angle is 35°.

Grade C

2 A radar station in Edinburgh has a range of 200 miles.
   A radar station in London has a range of 250 miles.
   London and Edinburgh are 400 miles apart.
   Sketch the area that the radar stations can cover.
   Use a scale of 1 cm = 200 miles.

•Edinburgh

•London

Systems of measurement

- The two main systems of measurement used in Britain are the **imperial system** and the **metric system**.

- The **imperial system** is based on measurements introduced many years ago. 12 inches = 1 foot, 16 ounces = 1 pound, 14 pounds = 1 stone.

- The **metric system** is used in Europe and most of the rest of the world. It is the standard system used in science and is based on the decimal system.

> **REMEMBER**
> Strictly speaking, what we refer to as *weight* should be called mass but the term *weight* is used in Foundation exams as it is in everyday use.

The metric system

- The basic unit of length is the **metre** (m). Other units are **millimetres** (mm), **centimetres** (cm) and **kilometres** (km).
  10 mm = 1 cm, 100 cm = 1 m, 1000 m = 1 km

- The basic unit of weight is the **kilogram** (kg). Other units are the **gram** (g) and the **tonne** (T).   1000 g = 1 kg, 1000 kg = 1 T

- The basic unit of capacity is the **litre** (l). Other units are **millilitres** (ml) and **centilitres** (cl).   1000 ml = 1 litre, 10 ml = 1 cl, 100 cl = 1 litre

> **REMEMBER**
> You should know the connection between the metric units.

The imperial system

- The basic unit of length is the **foot** (ft). Other units are **yards** (yd), **inches** (in) and **miles** (m). 12 in = 1 ft, 3 ft = 1 yd, 1760 yd = 1 m

- The basic unit of weight is the **pound** (lb). Other units are the **ounce** (oz), the **stone** (st) and the **ton** (ton).   16 oz = 1 lb, 14 lb = 1 st, 2240 lb = 1 ton

- The basic unit of capacity is the **pint** (pt). Other units are **gallons** (gall) and **quarts** (qt).   2 pt = 1 qt, 8 pt = 1 gall

> **REMEMBER**
> You do not need to know these conversions. If you need them in a question they will be given.

Conversions factors

- Because many imperial units, such as **miles** and **pounds**, are still in common use you need to be able to convert between them. To do this we use **conversion factors**.

- There are five conversion factors that you need to know:

  2.2 pounds ≈ 1 kilogram   1 foot ≈ 30 centimetres

  5 miles ≈ 8 kilometres   1.75 pints ≈ 1 litre   1 gallon ≈ 4.5 litres

> **REMEMBER**
> You need to know these five conversions. If you need any others in a question they will be given. However, another useful one to learn is 1 inch ≈ 2.5 centimetres. ≈ means 'approximately equal to'.

Questions

Grade G

1
  a Convert 120 centimetres to metres.
  b Convert 3500 grams to kilograms.
  c Convert 230 centilitres into litres.
  d Convert 45 millimetres into centimetres.

Grade F

2 Use a calculator to work out approximately:
  a how many pounds are equivalent to 25 kg
  b how many litres are equivalent to 8 gallons
  c how many cm are equivalent to 1 yard
  d how many miles are equivalent to 120 km.

Surface area and volume of 3-D shapes

Units of length, area and volume

- The basic unit of length is the **metre** (m). Other units are **centimetres** (cm) and **millimetres** (mm).

  10 mm = 1 cm, 100 cm = 1 m

- The basic unit of area is the **square metre** (m^2). Other units are **square centimetres** (cm^2) and **square millimetres** (mm^2).

  $100 \text{ mm}^2 = 1 \text{ cm}^2$, $10\ 000 \text{ cm}^2 = 1 \text{ m}^2$

- The basic unit of volume is the **cubic metre** (m^3). Other units are **cubic centimetres** (cm^3) and **cubic millimetres** (mm^3).

  $1000 \text{ mm}^3 = 1 \text{ cm}^3$, $1\ 000\ 000 \text{ cm}^3 = 1 \text{ m}^3$

- You need to be able to relate the areas and volumes of similar shapes.

> **REMEMBER**
> When converting between units of volume, such as cubic centimetres and cubic metres, a common mistake is just to divide by 100 instead of 1 000 000 (100^3). Learn the connections.

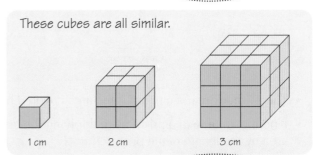

These cubes are all similar.

1 cm   2 cm   3 cm

The surface area of a cuboid

- A **cuboid** is a box shape, such as a cereal packet or a video cassette.

- The **surface area** of a cuboid is the **area** covered by its **net**.

> **REMEMBER**
> When working out the surface area of a cuboid each face occurs twice. You can use the formula:
> $SA = 2lw + 2lh + 2wh$

A cuboid measuring 1 cm by 2 cm by 3 cm has this net. Work out the surface area of the cuboid.

Two faces are 1 cm by 2 cm, two faces are 2 cm by 3 cm and two faces are 1 cm by 3 cm. This is calculated as:

$2 \times 1 \times 2 + 2 \times 2 \times 3 + 2 \times 1 \times 3 = 22 \text{ cm}^2$

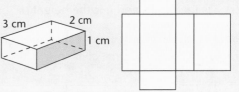
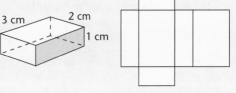

The volume of a cuboid

- The volume of a cuboid is simply the **product** of the **length**, **width** and **height**.

- The formula for the volume of a cuboid is: $V = lwh$

> **REMEMBER**
> In GCSE exams you should be given a diagram with a real-life Pythagoras' theorem problem. If not, draw one.

Work out the volume of a cuboid that is 3 cm by 4 cm by 6 cm.

$V = 3 \times 4 \times 6 = 72 \text{ cm}^3$

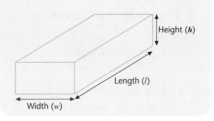

Height (*h*)

Length (*l*)

Width (*w*)

Questions

Grade B

1 a Convert 2 000 000 cm^3 to cubic metres.
  b Convert 30 000 cm^2 to square metres.
  c Convert 5 m^3 to cubic centimetres.
  d Convert 4 cm^2 to square millimetres

Grade B

2 Work out:
  a the surface area
  b the volume
  of a cuboid with sides of 2 cm, 3 cm and 4 cm.

Prisms

Prisms and cylinders

- A **prism** is a three-dimensional shape that always has the same **cross-section** when it is cut across, perpendicular to its length.

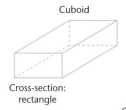

Cuboid

Cross-section: rectangle

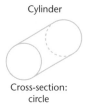

Triangular prism

Cross-section: isosceles triangle

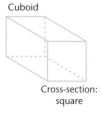

Cylinder

Cross-section: circle

Cuboid

Cross-section: square

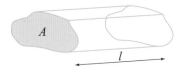

Hexagonal prism

Cross-section: regular hexagon

- The **surface area** of a prism is the area covered by its net.

- Find the **volume** of a prism by multiplying the cross-sectional area by the length of the prism. $V = Al$

> What is the volume of a prism with a cross-sectional area of 25 cm² and a length of 20 cm?
>
> Volume = 25 × 20 = 500 cm³

- A cylinder is a prism with a circular cross-section.

- The formula for the volume of a cylinder is: $V = \pi r^2 h$

  where r is the radius and h is the height or length of the cylinder.

> Find the volume of a cylinder with a radius 2 cm and height 10 cm. Give your answer in terms of π.
>
> $V = \pi r^2 h = \pi \times 2^2 \times 10 = 40\pi$ cm³

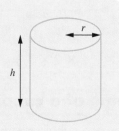

Questions

Grade C

PS 1 A cylinder has a height equal to its radius.
The volume is 393 cm³ to the nearest whole number.
What is the radius of the cylinder?

Grade C

2 A prism has a cross sectional area of 12 cm².
The volume of the prism is 33 cm³
What is the length of the prism?

Grade C

3 a A cuboid has dimensions 4 cm by 6 cm by 10 cm.
  i What is the cross-sectional area?
  ii What is the volume?

b A triangular prism has a cross-sectional area of 3.5 m² and a length of 12 m. What is its volume?

Grade C

4 What is the volume of a cylinder with a radius of 4 cm and a height of 10 cm? Give your answer in terms of π.

Pythagoras' theorem

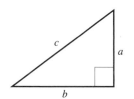

Pythagoras' theorem

- **Pythagoras' theorem** connects the sides of a **right-angled triangle**.

- Pythagoras' theorem states that: 'In any right-angled triangle, the square of the hypotenuse is equal to the sum of the squares of the other two sides.'

- The **hypotenuse** is the **longest** side of the triangle, which is always **opposite** the right angle.

- Pythagoras' theorem is usually expressed as a formula: $c^2 = a^2 + b^2$

  where c is the length of the hypotenuse and a and b are the lengths of the other two sides.

- The formula can be rearranged to find one of the other sides: $a^2 = c^2 - b^2$

> **REMEMBER**
> To find the actual value of a side, don't forget to take the square root.

Finding lengths of sides

- If you know the lengths of two sides of a right-angled triangle then you can always use Pythagoras' theorem to find the length of the third side.

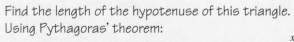

Find the length of the hypotenuse of this triangle.
Using Pythagoras' theorem:
$$x^2 = 9^2 + 6.2^2 = 81 + 38.44$$
$$= 119.44$$
$$x = \sqrt{119.44} = 10.9 \text{ cm}$$

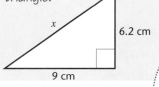

> **REMEMBER**
> Notice that when you are finding the **hypotenuse you add** the squares of the other two sides and when you are finding one of the **short sides, you subtract** the squares of the other two sides.

Find the length of the side marked y in this triangle.
Using Pythagoras' theorem:
$$y^2 = 16^2 - 12^2 = 256 - 144$$
$$= 112$$
$$y = \sqrt{112} = 10.6$$

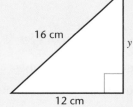

Real-life problems

- Pythagoras' theorem can be used to solve practical problems.

A ladder of length 5 m is placed with the foot 2.2 m from the base of a wall. How high up the wall does the ladder reach?
Using Pythagoras' theorem:
$$x^2 = 5^2 - 2.2^2$$
$$= 25 - 4.84$$
$$= 20.16$$
$$x = \sqrt{20.16} = 4.5 \text{ m}$$

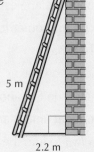

> **REMEMBER**
> In GCSE exams you should be given a diagram with a real-life Pythagoras' theorem problem. If not, draw one.

Questions

Grade C

1 A ladder is placed 1.5 m from the base of the wall and reaches 3.6 m up the wall. How long is the ladder?

Geometry grade booster

I can...

- [ ] find the perimeter of a 2-D shape
- [ ] find the area of a 2-D shape by counting squares
- [ ] draw lines of symmetry on basic 2-D shapes
- [ ] use the basic terminology associated with circles
- [ ] draw circles with a given radius
- [ ] recognise the net of a simple shape
- [ ] name basic 3-D solids
- [ ] recognise congruent shapes
- [ ] find the volume of a 3-D shape by counting squares

You are working at **Grade G** level.

- [ ] find the area of a rectangle, using the formula $A = lw$
- [ ] find the order of rotational symmetry for basic 2-D shapes
- [ ] measure and draw angles accurately
- [ ] use the facts that the angles on a straight line add up to 180° and the angles around a point add up to 360°
- [ ] draw and measure lines accurately
- [ ] draw the net of a simple 3-D shape
- [ ] read scales with a variety of divisions
- [ ] find the surface area of a 2-D shape by counting squares

You are working at **Grade F** level.

- [ ] find the area of a triangle using the formula $A = \frac{1}{2}bh$
- [ ] draw lines of symmetry on more complex 2-D shapes
- [ ] find the order of rotational symmetry for more complex 2-D shapes
- [ ] measure and draw bearings
- [ ] use the facts that the angles in a triangle add up to 180° and the angles in a quadrilateral add up to 360°
- [ ] find the exterior angle of a triangle and a quadrilateral
- [ ] recognise and find opposite angles
- [ ] draw simple shapes on an isometric grid
- [ ] tessellate a simple 2-D shape
- [ ] reflect a shape in the x- and y-axes
- [ ] convert from one metric unit to another
- [ ] convert from one imperial unit to another, given the conversion factor
- [ ] use the formula $V = lwh$ to find the volume of a cuboid
- [ ] find the surface area of a cuboid

You are working at **Grade E** level.

☐ find the area of a parallelogram, using the formula $A = bh$

☐ find the area of a trapezium, using the formula $\frac{1}{2}(a + b)h$

☐ find the area of a compound shape

☐ work out the formula for the perimeter, area or volume of simple shapes

☐ identify the planes of symmetry for 3-D shapes

☐ recognise and find alternate angles in parallel lines and a transversal

☐ recognise and find corresponding angles in parallel lines and a transversal

☐ recognise and find interior angles in parallel lines and a transversal

☐ use and recognise the properties of quadrilaterals

☐ find the exterior and interior angles of regular polygons

☐ understand the words 'sector' and 'segment' when used with circles

☐ calculate the circumference of a circle, giving the answer in terms of π if necessary

☐ calculate the area of a circle, giving the answer in terms of π if necessary

☐ recognise plan and elevation from isometric and other 3-D drawings

☐ translate a 2-D shape

☐ reflect a 2-D shape in lines of the form $y = a$, $x = b$

☐ rotate a 2-D shape about the origin

☐ enlarge a 2-D shape by a whole-number scale factor about the origin

☐ construct diagrams accurately, using compasses, a protractor and a straight edge

☐ use the appropriate conversion factors to change imperial units to metric units and vice versa

You are working at **Grade D** level.

☐ work out the formula for the perimeter, area or volume of complex shapes

☐ relate the exterior and interior angles in regular polygons to the number of sides

☐ find the area and perimeter of semicircles

☐ translate a 2-D shape, using a vector

☐ reflect a 2-D shape in the lines $y = x$, $y = -x$

☐ rotate a 2-D shape about any point

☐ enlarge a 2-D shape by a fractional scale factor

☐ enlarge a 2-D shape about any centre

☐ construct perpendicular and angle bisectors

☐ construct the perpendicular to a line from a point on the line and a point to a line

☐ draw simple loci

☐ work out the surface area and volume of a prism

☐ work out the volume of a cylinder, using the formula $V = \pi r^2 h$

☐ find the density of a 3-D shape

☐ find the hypotenuse of a right-angled triangle, using Pythagoras' theorem

☐ find the short side of a right-angled triangle, using Pythagoras' theorem

☐ use Pythagoras' theorem to solve real-life problems

You are working at **Grade C** level.

Geometry

Basic algebra

AQA 2/3   EDEXCEL 2/3   OCR 1/2

The language of algebra

> **REMEMBER**
>
> In the expression $6x$ there is an assumed times sign between 6 and x.

- **Algebra** is the use of **letters** instead of numbers to write rules, formulae and expressions.

- Algebra follows the same rules as arithmetic and uses the **same symbols** but there are some special rules in algebra.
  - 3 more than x is written as $x + 3$ or $3 + x$
  - 5 less than x is written as $x - 5$
  - p minus q is written as $p - q$
  - 8 times y is written as $8 \times y$ or $y \times 8$ or $8y$
  - b divided by 4 is written as $b \div 4$ or $\frac{b}{4}$
  - $1 \times n$ is just written as n
  - t times t is written as $t \times t$ or t^2

Simplifying expressions

- **Simplifying** an expression means writing it in as neat a form as possible.

- Simplify expressions by **collecting like terms** or **multiplying expressions**.

Collecting like terms

- To **collect like terms** combine terms that are similar, such as number terms or x-terms.

  a, $4a$ and $9a$ are like terms, $\frac{1}{2}x^2$ and $12x^2$ are like terms, $2ab$, $-3ba$ and $7ba$ are like terms.

- The number in front of the term is called the **coefficient**.

- **Group** all like terms together then **add** the **coefficients** in each group.

  To simplify $4a + 5b - 8 + 7a - 3b - 6$ rewrite as $4a + 7a + 5b - 3b - 8 - 6 = 11a + 2b - 14$.

Multiplying expressions

> **REMEMBER**
>
> Start by multiplying numbers and then multiply each letter, in alphabetical order. *Don't* get confused and *add* the numbers, which is a common error.

- To **multiply expressions**, multiply the numbers and use the index laws to multiply the letters.

  To simplify $4a^2b \times 3a^3b^2$ rewrite as $4 \times 3 \times a^2 \times a^3 \times b \times b^2 = 12a^5b^3$.

Questions

Grade F

1 Write the following as algebraic expressions.
- **a** p less than r
- **b** 7 plus x
- **c** a times b
- **d** t divided by 2
- **e** n more than m

Grade E

2 Simplify the following by collecting like terms.
- **a** $x + 2x + 5x$
- **b** $6x + 9 + 3x - 4$
- **c** $6w - 3k - 2w - 3k + 5w$
- **d** $9x^2 + 6z - 8z - 7x^2$

Grade D

3 Simplify the following by multiplying the expressions.
- **a** $3 \times 4t$
- **b** $4n^2 \times 3n$
- **c** $5mn \times 6m$
- **d** $-3x^2y \times 4xy^3$

Grade D

AU 4
- **a** Write an expression that simplifies to $5x$.
- **b** Write an expression that simplifies to $3a + 2b$.
- **c** Write an expression that simplifies to $6xy^2$.

Expanding and factorising

Expanding brackets

- **Expand** in mathematics means **multiply out**.

  Expressions such as $4(z + 3)$ and $5(x - 8)$ can be multiplied out.

- There is an invisible multiplication sign between the outside term and the opening bracket.

  $4(2x + 3)$ means $4 \times (2x + 3)$

- When expanding brackets, it is important to remember that the term outside the brackets is **multiplied** by each term inside the brackets.

  $4(2x + 3)$ means $4 \times (2x + 3) = 4 \times 2x + 4 \times 3 = 8x + 12$

  You would normally just write $5(2x - 7) = 10x - 35$.

> **REMEMBER**
> There is no need to show all the steps when expanding brackets as, usually, there is a mark for each correct term.

Expand and simplify

- When you are asked to **expand and simplify** an expression, it means expanding any brackets and then simplifying by **collecting like terms**.

  Expand and simplify $4(3 + m) - 5(2 - 3m)$.

  First, expand both brackets: $12 + 4m - 10 + 15m$

  Second, simplify: $2 - 11m$

> **REMEMBER**
> Be careful when expanding $-5 \times -3m = 15m$. Remember that $- \times -$ gives a plus answer.

> **REMEMBER**
> Do not try to expand and simplify in one go. If you try to do two things at once you will probably do one of them wrong.

Factorising

- **Factorisation** is the opposite of expanding brackets.

- **Factorisation** puts an expression back into the form $4(3x - 2)$.

- To factorise expressions, look for the **highest common factor** in each term.

  $5x + 20$ has a common factor of 5 in each term,

  so $5x + 20 = 5 \times x + 5 \times 4 = 5(x + 4)$

  $4xy - 8x^2$ has a common factor of $4x$

  so $4xy - 8x^2 = 4x \times y - 4x \times 2x = 4x(y - 2x)$

> **REMEMBER**
> Check your factorisation by multiplying out the final answer to see if it goes back to what you started with.

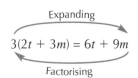

Expanding

$3(2t + 3m) = 6t + 9m$

Factorising

Questions

1 Expand the following.

  a $3(x + 5)$ **b** $5(y - 2)$

  c $3(2x + y)$ **d** $n(n - 7)$

  e $5m(2m + 3)$ **f** $3p^2(2p - 3q)$

2 Expand and simplify.

  a $2(x - 3) + 4(x + 3)$ **b** $5(m + 2) - 2(m - 6)$

  c $n(n + 1) + 3n(n - 2)$ **d** $6(x + 5) - 3(x + 1)$

  e $3x(2x - 3y) - 2x(x - y)$ **f** $8(x + 3y) + 2(x + 7y)$

3 Factorise the following expressions.

  a $5n + 10m$ **b** $6x^2 - 9x$

  c $5mn + 6m$ **d** $4x^2y + 12xy^2$

  e $2xy + 6x^2$ **f** $2a^2b - 8ab + 6ab^2$

4 Expand and simplify the following expressions, then factorise the answers.

  a $3(x + 2) + 2(6x - 9)$ **b** $5(2x + 3) - 5(x + 1)$

  c $4(x + 2) + 2(x - 6)$ **d** $3(x + 8) - 6(x + 1)$

  e $6(x + 1) + 9(x - 2)$ **f** $4(x + 7) + 6(x + 1)$

Linear equations

Solving linear equations

- An **equation** is formed when an expression is put equal to a number or another expression.
- A **linear equation** is one that only involves one **variable**.

  $2x + 3 = 7, 5x + 8 = 3x - 2$ are linear equations.

- **Solving** an equation means finding the value of the variable that makes it true.

  Solve $2x + 3 = 7$.

  The value of x that makes this true is 2 because $2 \times 2 + 3 = 7$

- The four ways to solve equations are shown below. These are all basically the same, but the most efficient is **rearrangement**.

Rearrangement

Solve $6x + 5 = 14$.

Move the 5 across the equals sign to give:

$6x = 14 - 5 = 9$

Divide both sides by 6 to give: $x = \frac{9}{6} \Rightarrow x = 1\frac{1}{2}$

Solve $\frac{y - 5}{3} = 6$.

Multiply both sides by 3 to give:

$y - 5 = 3 \times 6 = 18$

Move the 5 across the equals sign to give: $y = 18 + 5 = 23$

> **REMEMBER**
> This is called 'change sides, change signs', which means that plus becomes minus (and vice versa) and multiplication becomes division (and vice versa).

Inverse operations

Use the inverse operation method on simple equations such as $x + 7 = 9$ or $2x = 12$.

To solve the equation the opposite operation is applied to the right-hand side.

The inverse operation of $+7$ is -7 and of $2 \times x$ is 'divide by 2'.

$x + 7 = 9 \Rightarrow x = 9 - 7 = 2$

$2x = 12 \Rightarrow x = 12 \div 2 = 6$

Doing the same thing to both sides

Solve $3x - 5 = 16$.

Add 5 to both sides: $3x - 5 + 5 = 16 + 5$

$3x = 21$

Divide both sides by 3: $3x \div 3 = 21 \div 3$

$x = 7$

Inverse flow diagram

Solve $3x - 4 = 11$.

Flow diagram:

Inverse flow diagram:

Put through:

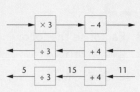

So the answer is 5.

> **REMEMBER**
> Always check that your answer works in the original equation.

Questions

Grade C

1 Solve the following linear equations.

  a $3x - 5 = 4$ **b** $2m + 8 = 6$

  c $\frac{n}{3} = 4$ **d** $8x + 5 = 9$

  e $\frac{x - 3}{5} = 2$ **f** $5x - 3 = 7$

  g $\frac{y + 2}{5} = 3$ **h** $\frac{x}{7} - 3 = 1$

  i $8 - 2x = 7$

Algebra

Solving equations with brackets

- When an equation contains brackets you must first multiply out the brackets and then solve the equation in the normal way.

Solve $4(x + 3) = 30$.

Multiply out: $4x + 12 = 30$

Subtract 12: $4x = 18$

Divide by 4: $x = 4\frac{1}{2}$

Equations with the variable on both sides of the equals sign

- When a letter appears on **both sides** of an equation use the 'change sides, change signs' rule.

- Use the rule to collect all the terms containing the **variable** on the **left-hand side** of the equals signs and all **number terms** on the **right-hand side**.

- When terms move across the equals sign, the signs change.

Rearranging $5x - 3 = 2x + 12$ gives $5x - 2x = 12 + 3$.

REMEMBER

Be careful when moving terms across the equals sign and remember to change the signs from plus to minus and vice versa.

- After the equation is rearranged, collect the terms together and solve the equation in the usual way.

$5x - 3 = 2x + 12$ is rearranged to $5x - 2x = 12 + 3 \Rightarrow 3x = 15 \Rightarrow x = 5$

Equations with brackets and the variable on both sides

- When an equation contains brackets and variables on **both sides**, always expand the brackets first.

Expanding $4(x - 2) = 2(x + 3)$ gives $4x - 8 = 2x + 6$.

- After expanding the brackets, rearrange the equation, collect the terms and solve the equation in the usual way.

$4(x - 2) = 2(x + 3)$ is expanded to $4x - 8 = 2x + 6 \Rightarrow 4x - 2x = 8 + 6$
$\Rightarrow 2x = 14$
$\Rightarrow x = 7$

Questions

Grade D

1 Solve the following by any method you choose.

  a $3(x - 5) = 12$ **b** $2(m - 3) = 6$

  c $3(x + 5) = 9$ **d** $2(x - 3) = 2$

  e $6(x - 1) = 9$ **f** $4(y + 2) = 2$

Grade D

2 Solve the following by any method you choose.

  a $3x - 2 = x + 12$ **b** $5y + 5 = 2y + 11$

  c $8x - 1 = 5x + 8$ **d** $7x - 3 = 2x - 8$

  e $6x - 2 = x - 1$ **f** $3y + 7 = y + 2$

Grade C

3 Solve the following by any method you choose.

  a $5(x - 3) = 2x + 12$ **b** $6(x + 1) = 2(x - 3)$

  c $3(x + 5) = 2(x + 10)$ **d** $7(x - 2) = 3(x + 4)$

Grade C

4 Solve the following by any method you choose.

  a $\dfrac{x + 1}{3} = 7$ **b** $\dfrac{x - 4}{5} = 2$

  c $\dfrac{4(x + 1)}{3} = 8$ **d** $\dfrac{2(x - 3)}{4} = 1$

Algebra

Setting up equations

• Linear equations can be used to **model** many practical problems.

The angles in a triangle are given as $2x$, $3x$ and $4x$.

What is the largest angle in the triangle?

The angles in a triangle add up to 180°, so:
$2x + 3x + 4x = 180°$

This simplifies to: $9x = 180°$

Divide by 9: $x = 20°$

So the largest angle is $4x = 4 \times 20 = 80°$.

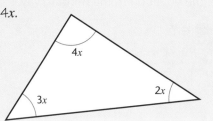

> **REMEMBER**
> Use the letter x when setting up an equation unless you are given another letter to use.

Fred is 29 years older than his daughter, Freda. Together their ages add up to 47.

Using the letter x to represent Freda's age, write down an expression for Fred's age.

Set up an equation in x and solve it to find Freda's age.

Freda's age is x.

Fred's age is $x + 29$.

The sum of both their ages is $x + x + 29 = 47$.

Collecting terms:
$2x + 29 = 47$

Subtracting 29 from both sides:
$2x = 18$

Dividing by 2:
$x = 9$

Questions

Grade D

1 The diagram shows a rectangle.

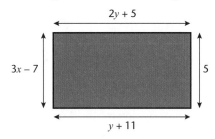

a What is the value of x?

b What is the value of y?

Grade D

PS 2 A family has x bottles of milk delivered every day from Monday to Friday and seven bottles delivered on Saturday. There is no delivery on Sunday.

In total they have 22 bottles delivered each week. Find the value of x.

Grade C

3 Asil thought of a number. He divided it by 2 then added 7. The result was 6 more than the number he first thought of.

a Use the information above to set up an equation, using x to represent the number Asil thought of.

b Solve your equation to find the value of x.

Algebra

Trial and improvement

Trial and improvement

- Some equations **cannot be solved** simply by algebraic methods.

- A **numerical method** for solving these equations is called **trial and improvement**.

- After an initial guess (**the trial**), an answer is calculated and compared to the required answer.

- A better guess (**the improvement**) is then made.

- This process is **repeated** until the answer is within a given accuracy.

- Normally the first thing to do is find **two whole numbers** between which the answer lies.

- Then find **two decimal numbers, each with one decimal place**, between which the answer lies.

- Then test the **midway value** to see which of the two numbers is closer.

> **REMEMBER**
>
> In GCSE examinations one or two initial guesses are always given to give you a start.

> Solve the equation $x^3 + 2x = 52$, giving your answer to one decimal place.
>
> Start with a guess of $x = 4$: $4^3 + 2 \times 4 = 64 + 8 = 72$ This is too high.
>
> Now try $x = 3$: $3^3 + 2 \times 3 = 27 + 6 = 33$ This is too low.
>
> The answer must be between 3 and 4.
>
> Now try $x = 3.5$: $3.5^3 + 2 \times 3.5 = 49.875$ This is close but too low.
>
> Now try $x = 3.6$: $3.6^3 + 2 \times 3.6 = 53.856$ This is close but too high.
>
> The answer must be between 3.5 and 3.6.
>
> To see which is closer, try $x = 3.55$:
>
> $$3.55^3 + 2 \times 3.55 = 51.838\ 875$$
>
> which is just too low.
>
> Hence, the answer, correct to one decimal place, is 3.6.

- Set out the working in a table.

Guess	$x^3 + 2x$	Comment
4	72	Too high
3	33	Too low
3.5	49.875	Too low
3.6	53.856	Too high
3.55	51.838875	Too low

> **REMEMBER**
>
> You must test the 'halfway' value between the one-decimal-place values.

Questions

Grade C

1 Use trial and improvement to find the solution to $x^3 - 2x = 100$.

Give your answer correct to one decimal place.

Grade C

2 Use trial and improvement to find the solution to $x^3 + x = 20$.

Give your answer correct to one decimal place.

Algebra

Formulae

Substitution

- Formulae in mathematics are expressed algebraically. The area of a rectangle is $A = lb$. To use a formula to work out a value, such as an area, **substitute** numbers into the formula.

- Substitution means **replacing letters** in formulae and expressions **with numbers**.

- When replacing letters with numbers, use brackets to avoid problems with minus signs.

  Work out the value of $ab + c$ if $a = -3, b = 4$ and $c = 5$. $ab + c = (-3)(4) + (5) = -12 + 5 = -7$

- Calculators have brackets keys that you use with more complicated expressions.

Formulae, expressions and equations

- **Formulae** (singular *formula*) are important in mathematics as they express many of the rules we use in a concise manner.

  The area of a rectangle: $A = lb$

- An **expression** is any arrangement of letters and numbers. The separate parts of expressions are called **terms**.

  $4a + 5b$ and $6x^2 - 2$ are expressions, $4a, 5b, 6x^2$ and -2 are terms.

- An **equation** is an expression that is put equal to another expression or a number.

- Equations can be **solved** to find the value of the variable that makes them true.

  $3x + 9 = 8$ and $4x + 7 = 2x - 1$ are equations.

> **REMEMBER**
> Make sure you know the difference between formulae, expressions and equations. They could be tested in an examination.

> **REMEMBER**
> Remember the rules, 'Change sides, change signs' and 'What you do to one side you must do to the other.'

Rearranging formulae

- The **subject** of a formula is the variable (letter) in the formula that stands on its own, usually on the left-hand side.

  x is the subject of each of these formula:
  $x = 5t + 4$ $x = 4(2y - 7)$ $x = \dfrac{1}{t}$

- To change the subject of a formula, **rearrange** the formula to get the new variable on the left-hand side.

- To rearrange formulae, use the same rules as for solving equations.

- Unlike an equation, at each step the right-hand side is not a number but an expression.

  Make m the subject of $y = m + 3$.
  Subtract 3 from both sides:
  $y - 3 = m$
  Reverse the formula:
  $m = y - 3$

Questions

Grade E

1 Which of the following are formulae, which are expressions and which are equations?

   a $V = lwh$ **b** $3x + 8 = 7$ **c** $P = 2l + 2w$
   d $(x - 1)(x + 1) = x^2 - 1$ **e** $4(x - 3) = 7$
   f $4x + 2y$ **g** $3(x + 1)^2$ **h** $c^2 = a^2 + b^2$

Grade D

2 Using $x = 3, y = -4$ and $z = 5$, work out the value of:

   a $2x + 3y$ **b** $x^2 - yz$ **c** $x(3y + 4z)$

Grade C

3 Rearrange each of the following formulae to make x the subject.

   a $T = 4x$ **b** $y = 2x + 3$ **c** $P = t + x$
   d $y = \dfrac{x}{5}$ **e** $A = mx$ **f** $S = 2\pi x$

Inequalities

AQA 3   EDEXCEL 3   OCR 2

Inequalities

- An **inequality** is an algebraic expression that uses the signs < (less than), > (greater than), ⩽ (less than or equal to) and ⩾ (greater than or equal to).

- The solution to an inequality is a **range of values**.

> The expression $x < 2$ means that x can take any value less than 2, all the way to minus infinity. x can also be very close to 2, for example, 1.9999… but is never actually 2.
>
> $x \geqslant 3$ means that x can take the value of 3 itself or any value greater than 3, up to infinity.

Solving inequalities

- **Linear inequalities** can be solved by the same rules that you use to solve equations.

- The answer when a linear inequality is solved is an inequality such as $x > -1$.

> Solve $2x + 3 < 11$.
>
> Solve $\frac{x}{5} - 3 \geqslant 4$.
>
> $2x < 11 - 3 \Rightarrow 2x < 8 \Rightarrow x < 4$ $\frac{x}{5} \geqslant 7 \Rightarrow x \geqslant 35$

> **REMEMBER**
>
> Don't use equal signs when solving inequalities as this doesn't get any marks in an examination.

Inequalities on number lines

- The solution to a linear inequality can be shown on a **number line**.

- Use the convention that an open circle is a **strict inequality** and a filled-in circle is an **inclusive inequality**.

> **REMEMBER**
>
> Inequalities often ask for integer values. An integer is a positive or negative whole number, including zero.

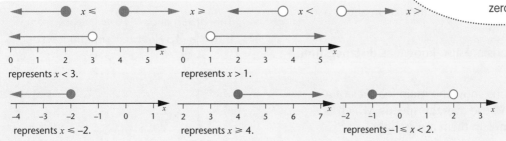

represents $x < 3$.

represents $x > 1$.

represents $x \leqslant -2$.

represents $x \geqslant 4$.

represents $-1 \leqslant x < 2$.

> Solve the inequality $2x + 3 < 11$ and show the solution on a number line.
>
> The solution is $x < 4$, which is shown on this number line.
>
>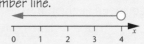

Questions

Grade C

1 Solve the following inequalities.

   a $x + 5 < 8$ **b** $2x + 3 > 5$ **c** $\frac{x}{3} - 5 \geqslant 1$

   d $4x + 6 \leqslant 2$ **e** $\frac{x}{2} + 7 > 2$ **f** $3x + 8 \leqslant 5$

Grade C

2 What inequalities are shown by the following number lines?

   a

   b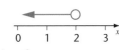

Grade C

3 a What inequality is shown on this number line?

   b Solve the inequality $3x + 6 \geqslant 3$.

   c What integers satisfy both of the inequalities in parts (**a**) and (**b**)?

Algebra

Conversion graphs

- A conversion graph is a straight line graph that shows a relationship between two variables.

Graph 1 shows the relationship between litres and gallons.

Graph 2 shows the charge for units of electricity.

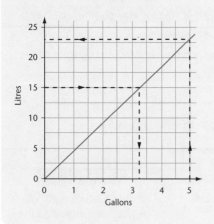

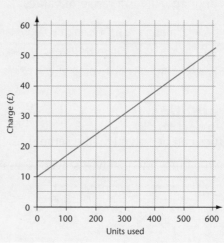

- Conversion graphs can be used to convert values from one unit to another.

In graph 1, 'litres and gallons', 15 litres is shown to be about $3\frac{1}{4}$ gallons and 5 gallons is shown to be about 23 litres.

REMEMBER

When asked for an average speed, answer in kilometres per hour (km/h) or miles per hour (mph).

Travel graphs

- A **travel graph** shows information about a journey.

- Travel graphs are also known as **distance-time** graphs.

- Travel graphs show the main features of a journey and use **average speeds**, which is why the lines in them are straight.

- In **reality**, vehicles do not travel at constant speeds.

- The average speed is given by:

  average speed = $\dfrac{\text{total distance travelled}}{\text{total time taken}}$

- In a travel graph, the steeper the line, the **faster** the vehicle is travelling.

This graph shows a car journey from Barnsley to Nottingham and back.

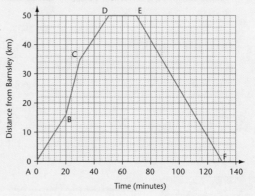

Questions

Grade E

1. Refer to graph 2, above, showing the relationship between cost and units of electricity.

   a How much will a customer who uses 500 units be charged?

   b How many units will a customer who is charged £20 have used?

Grade D

2. Refer to the travel graph, above.

   a After how many minutes was the car 16 kilometres from Barnsley?

   b What happened between points D and E?

   c During which part of the journey was the car travelling fastest?

   d What was the average speed for the part of the journey between C and D?

Linear graphs

Negative coordinates

- You have already met **negative coordinates** and they often occur when drawing graphs.

The coordinates of A are (1, 2), those of E are (−3, 2) and J is (4, −2).

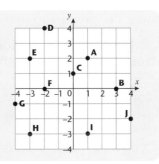

E

Drawing graphs from tables

The table shows values of the function $y = 2x + 1$ for values of x from −3 to +3.

Use the table to draw the graph of $y = 2x + 1$.

x	−3	−2	−1	0	1	2	3
y	−5	−3	−1	1	3	5	7

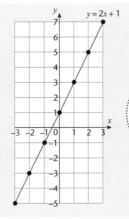

REMEMBER

Always label graphs. If you draw two graphs on the same set of axes and do not label them you will lose marks.

D

Drawing linear graphs

- You only need two points to draw a straight line.
- It is better to use three points, as the third point acts as a check.

Draw the graph of $y = 3x - 1$.

Pick values for x, such as $x = 3$, and work out the equivalent y-values.

$x = 3 \Rightarrow y = 3 \times 3 - 1 = 8$ giving (3, 8)

$x = 1 \Rightarrow y = 3 \times 1 - 1 = 2$ giving (1, 2)

$x = 0 \Rightarrow y = 3 \times 0 - 1 = -1$ giving (0, −1).

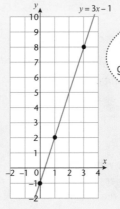

REMEMBER

In examinations you are always given a grid and told the range of x-values to use.

REMEMBER

Always use $x = 0$ as a point as it makes the calculation easy.

D

Questions

Grade E

1 Refer to the grid, above. Write down the coordinates of B, C, D, F, G, H and I.

Grade E

2 a Complete the table of values for the function $y = 2x - 3$ for values of x from −3 to +3.

x	−3	−2	−1	0	1	2	3
y	−9	−7				1	3

b Draw a set of axes with x-values from −3 to +3 and y-values from −9 to +3. Use the table to draw the graph of $y = 2x - 3$.

Grade D

3 Draw a set of axes with x-values from −3 to +3 and y-values from −11 to +13.

Draw the graph of $y = 4x + 1$ for values of x from −3 to +3.

Algebra

Gradient-intercept

Gradients

- The gradient of a line is a measure of how **steep** the line is.

- It is calculated by **dividing** the **vertical** distance between two points on the line by the **horizontal** distance between the same two points.

$$\text{gradient} = \frac{\text{vertical distance}}{\text{horizontal distance}}$$

- This is sometimes written as $\text{gradient} = \dfrac{y\text{-step}}{x\text{-step}}$

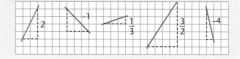

These lines have gradients as shown.

- Lines that slope down from left to right have **negative** gradients.

- The **right-angled triangles** drawn along grid lines are used to find gradients.

- To draw a line with a certain **gradient**, for every unit moved **horizontally**, move upwards (or downwards if the gradient is negative) by the number of units of the gradient.

Draw lines with gradients of $\frac{1}{4}$ and -2.

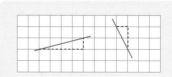

The gradient-intercept method

- The **gradient–intercept** method is the easiest and quickest method for drawing graphs.

- In the function $y = 2x + 3$, the **coefficient** of x (2) is the **gradient** and the **constant** term $(+3)$ is the **intercept**.

- The intercept is the point where the line **crosses the y-axis**. (See the example below.)

Drawing a line with a given gradient

Draw the line $y = 2x + 3$.

Start by marking the intercept point $(0, 3)$.

Next, move 1 unit across and 2 units up to show the gradient.

Repeat this a few more times.

Join up the points to get the required line.

> **REMEMBER**
> Draw graphs with a sharp pencil and use a ruler to make sure the lines are straight.

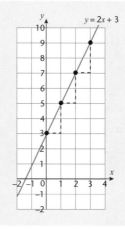

Questions

Grade C

AU 1 Here are the equations of four lines.

A: $y = 2x - 3$ B: $y = 3x - 3$
C: $y = 2x + 1$ D: $y = \frac{1}{2}x - 1$

a Which two lines are parallel?

b Which two lines cross the y-axis at the same point?

Grade C

2 a Draw a set of axes with x-values from -3 to $+3$ and y-values from -9 to $+15$. On these axes, draw the graph of $y = 4x + 3$.

b Draw a set of axes with x-values from -4 to $+4$ and y-values from -1 to $+3$. On these axes, draw the graph of $y = \frac{1}{2}x + 1$.

Algebra

Quadratic graphs

Drawing quadratic graphs

A quadratic graph has an x^2 term in its equation.

$y = x^2$, $y = x^2 + 2x + 3$ will give quadratic graphs.

- Quadratic graphs always have the same characteristic shape, which is called a **parabola**.

REMEMBER

Try to draw a smooth curve through all the points. Examiners prefer a good attempt at a curve, rather than points joined with a ruler.

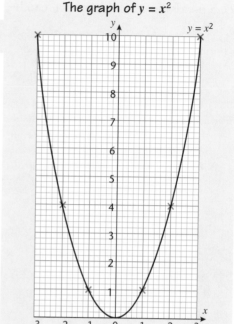

The graph of $y = x^2$

- Quadratic graphs are drawn from tables of values.

This table shows the values of $y = x^2 + 2x - 3$ for values of x from -4 to 2.

x	-4	-3	-2	-1	0	1	2
y	5	0	-3	-4	-3	0	5

The graph of $y = x^2 + 2x - 3$

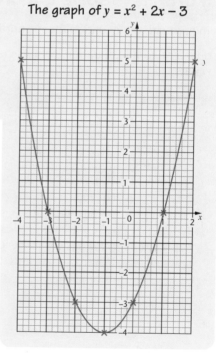

- Usually you will be asked to fill in some values in a table and then draw the graph.

The table shows some y-values for $y = x^2 - x - 1$.

x	-3	-2	-1	0	1	2	3
y	11	5			1		5

REMEMBER

As all parabolas have line symmetry, there will be some symmetry in the y-values in the table.

Questions

Grade C

1 a Complete the table of values above for $y = x^2 - x - 1$.

b Draw the graph of $y = x^2 - x - 1$.
Label the x-axis from 3 to $+3$ and the y-axis from -2 to 12.

Algebra

Reading values from quadratic graphs

- Once a quadratic graph is drawn it can be used to solve various equations.

Use the graph of $y = x^2 - 2$ to find the x-values when $y = 5$.

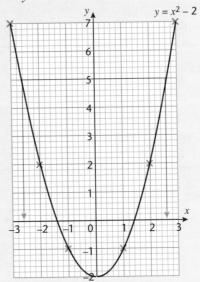

REMEMBER

Always show the lines and arrows because, even if you make an error drawing the graph, you can still get marks for reading values from your graph.

Draw the line $y = 5$ and draw down to the x-axis from the points where the line intercepts the curve.

The x-values are about $+2.6$ and -2.6.

Using graphs to solve quadratic equations

- Solving a quadratic equation means finding the x-values that make it true.

- To solve a quadratic equation from its graph, read the values where the curve crosses the x-axis.

Solve the equation $x^2 - 3x - 4 = 0$.

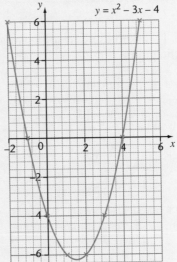

REMEMBER

The graph is
$y = x^2 - 3x - 4$ and the equation is
$x^2 - 3x - 4 = 0$.
The solution of the equation is found where the graph crosses the x-axis, where $y = 0$.

The graph crosses the x-axis at $x = -1$ and $x = 4$.

Questions

Grade C

1 a Use the graph of $y = x^2 - 2$ to solve the equation $x^2 - 2 = 0$.

b Use the graph of $y = x^2 - 3x - 4$ to find the x-values when $y = 2$.

Algebra

Pattern

Patterns in number

- There are many curious and interesting patterns formed by numbers in mathematics.

$0 \times 9 + 1 = 1$
$1 \times 9 + 2 = 11$
$12 \times 9 + 3 = 111$
$123 \times 9 + 4 = 1111$
$1234 \times 9 + 5 = 11111$

$1 \times 8 + 1 = 9$
$12 \times 8 + 2 = 98$
$123 \times 8 + 3 = 987$
$1234 \times 8 + 4 = 9876$
$12345 \times 8 + 5 = 98765$

REMEMBER
When studying patterns formed by repeated calculations look for ways in which parts of the calculation continue in sequences.

$1 \times 3 \times 37 = 111$
$2 \times 3 \times 37 = 222$
$3 \times 3 \times 37 = 333$
$4 \times 3 \times 37 = 444$

$7 \times 7 = 49$
$67 \times 67 = 4489$
$667 \times 667 = 444889$
$6667 \times 6667 = 44448889$

- Use a calculator to check that these work. If you can spot the patterns, it is possible to write down the next term without using a calculator.

- Spotting patterns is an important part of mathematics. It helps in solving problems and making calculations.

Number sequences

- A **number sequence** is a series of numbers that build up according to some **rule**.

- A **linear sequence** is one in which terms increase by a **constant difference**.

  $2, 8, 14, 20, 26, 32, \ldots$ is a linear sequence with an increase of $+6$ each time.

- Sometimes patterns build up according to more complicated rules.

  $2, 4, 8, 16, 32, 64, \ldots$ is formed by doubling the previous term.
  $2, 5, 11, 23, 47, \ldots$ is formed by multiplying the previous term by 2 and adding 1.
  $1, 1, 2, 3, 5, 8, 13, 21, \ldots$ is formed by adding the two previous terms.

- One way to spot how a pattern is increasing is to look at the **differences** between consecutive terms.

- You can use differences to predict the next terms.

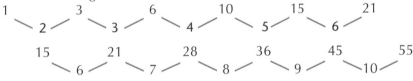

Questions

AU 1 Do not use a calculator for this question.

  a Write the next two lines of the first pattern, top left, above, after the line:
  $1234 \times 9 + 5 = 11111$

  b Write the next two lines of the second pattern, top right, above, after:
  $12345 \times 8 + 5 = 98765$

  c Write the next two lines of the third pattern, bottom left, above, after the line:
  $4 \times 3 \times 37 = 444$

  d Write the next two lines of the fourth pattern, bottom right, above, after the line:
  $6667 \times 6667 = 44448889$

2 For each of the following patterns:

  i write down the next two terms

  ii describe how the pattern is building up.

  a 3, 9, 15, 21, 27, 33, ...

  b 3, 6, 12, 24, 48, ...

  c 100, 96, 92, 88, 84, 80, ...

  d 3, 5, 9, 15, 23, 33, 45, ...

  e 16, 13, 10, 7, 4, 1, ...

Algebra

The nth term

The nth term of a sequence

- When using a number sequence, you may need to know the **50th** or **higher term**.
- This can be found by counting on but mistakes are likely, so it is easier to find the **nth term**.
- The nth term is an **algebraic expression** that gives any term in the sequence.

  The nth term of a sequence is given as:

  $An + b$

  where A, the coefficient of n, is the difference between one term and the next term (**consecutive** terms) and b is the difference between the first term and A.

- To find any term **substitute** a number for n.

  > The nth term of a sequence is found by the rule $3n + 1$.
  > What are the first three terms and the 40th term?
  > Take $n = 1$: $3 \times 1 + 1 = 4$
  > Take $n = 2$: $3 \times 2 + 1 = 7$
  > Take $n = 3$: $3 \times 3 + 1 = 10$
  > Take $n = 40$: $3 \times 40 + 1 = 121$
  > So the first three terms are 4, 7 and 10 and the 40th term is 121.

Finding the nth term

- The nth term of a linear sequence is always of the form $An \pm b$.

  > $3n + 1, 4n - 3, 8n + 7$ are typical nth terms.

- To find the coefficient of n find the constant difference of the sequence.

  > The sequence 4, 7, 10, 13, 16, 19, ... has a constant difference of 3, so the nth term of the sequence will be given by $3n \pm b$.

- To find the value of b, work out what the difference is between the coefficient of n and the first term of the sequence.

  > The sequence 4, 7, 10, 13, 16, 19, ... has a first term of 4. The coefficient of n is 3. To get from 3 to 4, add 1, so the nth term is $3n + 1$.
  >
  > What is the nth term of the sequence 4, 9, 14, 19, 24, 29, ...?
  > The constant difference is 5, and $5 - 1 = 4$, so the nth term is $5n - 1$.

Questions

Grade E

1 a The nth term of a sequence is $4n - 1$.

  i Write down the first three terms of the sequence.

  ii Write down the 100th term of the sequence.

 b The nth term of a sequence is $\frac{1}{2}(n + 1)(n + 2)$.

  i Write down the first three terms of the sequence.

  ii Write down the 199th term of the sequence.

Grade D

2 Write down the nth term of each of these sequences.

 a 6, 11, 16, 21, 26, 31, ...

 b 3, 11, 19, 27, 35, ...

 c 9, 12, 15, 18, 21, 24, ...

Grade C

AU 3 Two sequences are

2, 7, 12, 17, 22, ...   4, 14, 24, 34, 44, ...

Explain why the sequences will never have a term in common.

Algebra

Sequences

Special sequences

- There are many special sequences that you should be able to recognise.
  - The **even numbers**
    2, 4, 6, 8, 10, 12, 14, … The nth term is $2n$.
  - The **odd numbers**
    1, 3, 5, 7, 9, 11, 13, … The nth term is $2n - 1$.
  - The **square numbers**
    1, 4, 9, 16, 25, 36, 49, … The nth term is n^2.
  - The **triangular numbers**
    1, 3, 6, 10, 15, 21, 28, … The nth term is $\frac{1}{2}n(n + 1)$.
  - The **powers of 2**
    2, 4, 8, 16, 32, 64, 128, … The nth term is 2^n.
  - The **powers of 10**
    10, 100, 1000, 10 000, 100 000, … The nth term is 10^n.
  - The **prime numbers**
    2, 3, 5, 7, 11, 13, 17, 19, … There is no nth term as there is no
    pattern to the prime numbers.

> **REMEMBER**
> The only even prime number is 2

Finding the nth term from given patterns

- An important part of mathematics is to find **patterns in situations**.
- Once a **pattern** has been found the **nth term** can be used to describe the pattern.

The diagram shows a series of 'L' shapes.

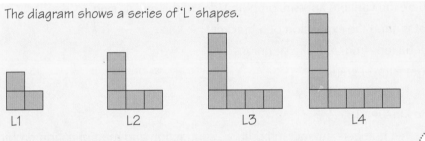

L1   L2   L3   L4

> **REMEMBER**
> Write down the sequence and look for the nth term.

How many squares are there in the 100th 'L' shape?
First write down the sequence of squares in the 'L' shapes.
3, 5, 7, 9, …
The constant difference is 2, and $2 + 1 = 3$, so the nth term is $2n + 1$.
The 100 term will be $2 \times 100 + 1 = 201$, so there will be 201 squares in L100.

Questions

Grade D

1 a What is the 100th even number?
   b What is the 100th odd number?
   c What is the 100th square number?
   d Continue the sequence of triangular numbers for the next five terms.
   e Continue the sequence of powers of 2 for the next five terms.
   f Write down the next five prime numbers after 19.

Grade C

2 Matches are used to make pentagonal patterns.

   1   2   3   4

   a How many matches will be needed to make the 10th pattern?
   b How many matches will be needed to make the nth pattern?

Algebra

Algebra grade booster

I can…
- ☐ use a formula expressed in words
- ☐ substitute numbers into expressions
- ☐ use letters to write a simple algebraic expression
- ☐ solve linear equations that require only one inverse operation to solve
- ☐ read values from a conversion graph
- ☐ plot coordinates in all four quadrants
- ☐ give the next value in a linear sequence
- ☐ describe how a linear sequence is building up

You are working at **Grade F** level.

- ☐ simplify an expression by collecting like terms
- ☐ simplify expressions by multiplying terms
- ☐ solve linear equations that require more than one inverse operation to solve
- ☐ read distances and times from a travel graph
- ☐ draw a linear graph from a table of values
- ☐ find any number term in a linear sequence
- ☐ recognise patterns in number calculations

You are working at **Grade E** level.

- ☐ use letters to write more complicated algebraic expressions
- ☐ expand expressions with brackets
- ☐ factorise simple expressions
- ☐ solve linear equations where the variable appears on both sides of the equals sign
- ☐ solve linear equations that require the expansion of a bracket
- ☐ set up and solve simple equations from real-life situations
- ☐ find the average speed from a travel graph
- ☐ draw a linear graph without a table of values
- ☐ substitute numbers into an nth term rule
- ☐ understand how odd and even numbers interact in addition, subtraction and multiplication problems

You are working at **Grade D** level.

- ☐ expand and simplify expressions involving brackets
- ☐ factorise expressions involving letters and numbers
- ☐ recognise an expression, an equation and a formula
- ☐ solve linear equations that have the variable on both sides and include brackets
- ☐ solve simple linear inequalities
- ☐ show inequalities on a number line
- ☐ solve equations using trial and improvement
- ☐ rearrange simple formulae
- ☐ use a table of values to draw a simple quadratic graph
- ☐ use a table of values to draw a more complex quadratic graph
- ☐ solve a quadratic equation from a graph
- ☐ give the nth term of a linear sequence
- ☐ give the nth term of a sequence of powers of 2 or 10

You are working at **Grade C** level.

Answers

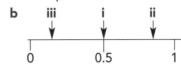

Page 16

1 a 28 and 18   b 1 tail and 1 head
  c 60
2 a 14 hours   b ❀ ❀ ❀ ✦

Page 17

1 a Thursday and Saturday
  b 144   c Friday
  d Wednesday, as most copies were sold.
2 a Town B, about 4 degrees hotter
  b Town B, summer is in December
  c September
  d No, the lines cross but they have no meaning, they just show trends.

Page 18

1 a 5   b Y
  c Because it is the smallest value.
2 a 7   b 8.5
3 a 64   b 35
4 2, 6, 8, 8, 11

Page 19

1 a 7   b 12
2 a Dan 76, Don 76   b Dan 125, Don 21
  c He could pick Dan as he can sometimes achieve high scores or he can pick Don because he is consistent and always achieves a reasonable score.
3 Asaf: median; Brian: mode; Clarrie: mean
4 Mode as there are five people with this wage.

Page 20

1 a 30   b 0
  c i 45   ii 1.5
2 a $20 < x \leqslant 30$ b i 1350   ii 27

Page 21

1 a 16, 18, 19, 23, 26, 29, 30, 31, 32, 35, 38, 42, 42, 42, 57

```
1 | 6 8 9
2 | 3 6 9
3 | 0 1 2 5 8
4 | 2 2 2
5 | 7
```
Key 4 | 2 = 42
  b 42   c It is an extreme value.
  d 31   e 41
  f 32
2 a 20   b i 13   ii 49   iii 36
  c 35   d 26.5
  e i 560   ii 28

Page 22

1 a i Certain   ii Evens   iii Very likely
    iv impossible
  b

```
        iii        i          ii
         ↓         ↓          ↓
    |_____|
    0               0.5              1
```

2 a $\frac{1}{2}$ b $\frac{1}{13}$
  c $\frac{1}{26}$
3 a $\frac{1}{6}$ b $\frac{12}{52} = \frac{3}{13}$
  c $\frac{1}{2}$

Page 23

1 $\frac{7}{10}$
2 a $\frac{4}{52} = \frac{1}{13}$ b $\frac{4}{52} = \frac{1}{13}$
  c $\frac{8}{52} = \frac{2}{13}$
3 a Ali 0.25 Barry 0.22 Clarrie 0.19
  b Clarrie, most trials c 40

Page 24

1 a 36
  b i $\frac{6}{36} = \frac{1}{6}$ ii $\frac{4}{36} = \frac{1}{9}$
  c i $\frac{4}{36} = \frac{1}{9}$ ii $\frac{6}{36} = \frac{1}{6}$ iii 7

Page 25

1 a 50   b 20
  c 70
2 2 red and 4 green
3 a i 5 ii 30 iii 13 iv $\frac{12}{30} = \frac{2}{5}$
  b i 200 ii 5% iii 40%

Page 26

1 a 120   b 33°
2 Pie chart with angle of 144° for Red, 60° for Blue, 36° for White, 72° for Black and 48° for Green with a title 'favourite colours' and each sector labelled.

Page 27

1 a The scatter diagram shows positive correlation.
  b Students who can learn their tables can usually also learn spellings.
2 a As a car gets older, its value decreases.
  b There is no relationship between someone's wages and the distance they live from work.
3 73

Page 28

1 **a** Leading question, two questions in one, double negative, not enough responses

b Overlapping responses, missing responses

2 For the outside courgettes the mean is 13.5 and the range is 9.

The mean of the greenhouse courgettes is 15.5 – 13.5 = 2 cm longer.

The range of the greenhouse courgettes is 3 cm less.

The data supports the hypothesis that the courgettes grown in the greenhouse are bigger, although the sample size is small. The greenhouse courgettes are also more consistent than the outside courgettes as the range is smaller.

Page 30

1 **a i** 24 **ii** 21 **iii** 45 **iv** 48 **v** 72 **vi** 49

b i 8 **ii** 7 **iii** 7 **iv** 3 **v** 9 **vi** 8

c i 5 r 5 **ii** 9 r 1 **iii** 7 r 1 **iv** 9 r 1 **v** 6 r 6 **vi** 6 r 2

2 **a i** 14 **ii** 6 **iii** 15 **iv** 13 **v** 16 **vi** 16

b i 27 **ii** 30 **iii** 11 **iv** 3 **v** 6 **vi** 1

c i $5 \times (6 + 1) = 35$ **ii** $18 \div (2 + 1) = 6$
iii $(25 - 10) \div 5 = 3$ **iv** $(20 + 12) \div 4 = 8$

Page 31

1 **a** 5 hundred

b Twenty-seven thousand, seven hundred and eight

c 2 406 502

2 **a i** 60 **ii** 140 **iii** 50

b i 700 **ii** 700 **iii** 1300

3 **a** 5830 **b** 2578

Page 32

1 **a i** 273 **ii** 324 **b i** 26 **ii** 34

c 84 **d** 30

2 **a** 543 **b** 107

c 496 **d** 52

3 37

Page 33

1 **a** $\frac{7}{15}$ **b** $\frac{6}{9}$ ($\frac{2}{3}$)

c $\frac{3}{8}$ **d** $\frac{4}{6}$ ($\frac{2}{3}$)

e $\frac{4}{9}$

2 **a i** $\frac{5}{11}$ **ii** $\frac{3}{5}$ **iii** $\frac{6}{8}$ ($\frac{3}{4}$)

b i $\frac{6}{9}$ ($\frac{2}{3}$) **ii** $\frac{3}{5}$ **iii** $\frac{2}{13}$

3 **a** $\frac{3}{6}$, $\frac{5}{10}$ and $\frac{20}{40}$

b $\frac{4}{10}$, $\frac{8}{20}$ and $\frac{40}{100}$

c b and d

4 $\frac{1}{2}$ has not got a denominator of 8.

$\frac{4}{8}$ is the only one that can be cancelled down

$\frac{1}{8}$ is the only one not equal to a half.

Page 34

1 **a i** 18 **ii** 24 **iii** 20

b i $\frac{2}{5}$ **ii** $\frac{1}{5}$ **iii** $\frac{3}{10}$ **iv** $\frac{3}{5}$ **v** $\frac{5}{7}$

c Any valid fractions such as **i** $\frac{12}{16}$ **ii** $\frac{3}{18}$ **iii** $\frac{6}{16}$

2 **a i** $2\frac{2}{5}$ **ii** $3\frac{1}{4}$ **iii** $2\frac{2}{7}$ **iv** $2\frac{5}{8}$ **v** $5\frac{2}{3}$

b i $\frac{8}{3}$ **ii** $\frac{23}{5}$ **iii** $\frac{20}{7}$ **iv** $\frac{9}{4}$ **v** $\frac{29}{8}$

Page 35

1 **a i** $\frac{4}{5}$ **ii** $\frac{2}{7}$ **iii** $\frac{5}{9}$ **iv** $\frac{1}{3}$

b i $\frac{1}{3}$ **ii** $\frac{1}{5}$ **iii** $\frac{2}{3}$ **iv** $\frac{1}{3}$

c i $1\frac{4}{9}$ **ii** $1\frac{2}{13}$ **iii** $1\frac{4}{11}$ **iv** $1\frac{1}{3}$

d i $1\frac{1}{2}$ **ii** $1\frac{3}{5}$ **iii** $1\frac{1}{3}$ **iv** $1\frac{1}{2}$

2 **a** 8 **b** 20

c 25 **d** £225

e 18 kg **f** 4 hours

g $\frac{3}{4}$ of 20 = 15 ($\frac{2}{3}$ of 21 = 14)

h $\frac{4}{7}$ of 63 = 36 ($\frac{7}{8}$ of 40 = 35)

Page 36

1 **a** $\frac{1}{5}$ **b** $\frac{3}{8}$

c $\frac{2}{9}$ **d** $\frac{3}{5}$

e $\frac{1}{36}$ **f** $\frac{1}{10}$

g $\frac{1}{6}$ **h** $\frac{5}{12}$

2 **a** $\frac{2}{5}$ **b** $\frac{3}{4}$

c $\frac{1}{3}$

3 **a** $\frac{1}{3}$ **b** $\frac{1}{8}$

c £480

4 Jane is 16, John is 14, so two years older.

Page 37

1 a 0.375   b 0.4
   c 0.45   d 0.2222...
   e 0.531 25   f 0.454 545...
2 a $\frac{7}{11}, \frac{16}{25}, \frac{13}{20}, \frac{2}{3}$ b $\frac{12}{25}, \frac{5}{11}, \frac{9}{20}, \frac{4}{9}$
3 a $0.3\dot{6}$ b $0.6\dot{1}\dot{5}$
   c $0.3\dot{6}$ d $0.\dot{6}$
   e $0.16\dot{6}$ f $0.\dot{7}$
4 a $\frac{4}{5}$ b $\frac{13}{20}$
   c $\frac{1}{8}$ d $2\frac{9}{20}$
   e $\frac{1}{40}$ f $\frac{111}{125}$

Page 38

1 a −£7   b −40 m
   c +8   d 24°F
   e +50 km   f +2 km
2 a i < ii > iii <   b −1, −4
   c i 5 ii −3 iii −3.5
3 −3 and −6

Page 39

1 a −4   b −12
   c 2   d −3
   e 0   f −4
   g −6   h 5
   i −14   j −4
   k 12   l −8
   m 7   n −22
   o −8
2 a −10   b −24
   c 21   d 6
3 a −7 + 14 = +7   b 9 − 17 = −8
   c −6 + 14 − 5 = 3
4 a Any values such as 6 and 8.
   b Any values such as −4 and −2.

Page 40

1 a i 6, 12, 18, 24, 30   ii 13, 26, 39, 52, 65
     iii 25, 50, 75, 100, 125
   b i 250, 62, 78, 90, 108, 144, 96, 120
     ii 78, 90, 108, 144, 81, 96, 120, 333
     iii 250, 90, 85, 35, 120, 125
     iv 90, 108, 144, 81, 333 v 250, 90, 120
   c Yes because 8 + 1 + 9 = 18 = 2 × 9

2 a i {1, 2, 3, 4, 6, 8, 12, 24} ii {1, 3, 5, 15}
     iii {1, 2, 5, 10, 25, 50}
     iv {1, 2, 4, 5, 8, 10, 20, 40}
   b {1, 2, 3, 4, 6, 8, 9, 12, 16, 18, 24, 36, 48, 72, 144}

Page 41

1 a i {1, 2, 3, 6, 9, 18} ii {1, 19}
     iii {1, 2, 4, 5, 10, 20}
   b 19
   c 61, 79, 83, 17, 41, 29
   d 2
2 a 36, 49, 64, 81, 100
   b i 121 ii 144 iii 169 iv 196 v 225
   c   1 + 3 + 5 + 7 + 9 = 25
        1 + 3 + 5 + 7 + 9 + 11 = 36
     1 + 3 + 5 + 7 + 9 + 11 + 13 = 49

Page 42

1 a i 9 ii 8 iii 5
   b i +2, −2 ii +4, −4 iii +10, −10
   c i 24 ii 2.5 iii 6.1
   d i 125 ii 1 iii 1000
   e i 1.2 ii 16 iii 0.3
2 a i 27 ii 64 iii 1000
   b i 4^5 ii 6^6 iii 10^4 iv 2^7
   c i 1024 ii 46 656 iii 10 000
     iv 128
   d 64, 128, 256, 512, 1024

Page 43

1 a 800   b 64
   c 250   d 300
   e 760   f 3250
   g 6.4   h 0.028
   i 3.9   j 0.075
   k 0.034   l 0.94
2 a 600 000   b 12 000
   c 140 000   d 200
   e 20   f 50

Answers

Page 44

1 a i $2^3 \times 3^2$ ii $2^2 \times 3 \times 5^3$ iii $3^2 \times 5^2$
   iv $2 \times 3^3 \times 7$
   b i 72 ii 1500 iii 225 iv 378
2 a 30   b 84
   c 130   d 36
   e 40   f 300
3 a $2^2 \times 5$ b $3^2 \times 5$
   c 2^6 d $2^3 \times 3 \times 5$
4 a 2^4 b $2 \times 3 \times 7$
   c $2 \times 5 \times 7$ d $2^3 \times 5^2$

Page 45

1 a i 30 ii 21 iii 39
   b The LCM is the product of the two numbers.
   c i 18 ii 40 iii 75
2 a 6   b 2
   c 5   d 16
   e 12   f 1
3 a 60   b 150
   c 120

Page 46

1 a i 2^7 ii 2^9 iii 2^6
   b i 3^3 ii 3^4 iii 3^3
   c i x^9 ii x^9 iii x^{11}
   d i x^4 ii x^6 iii x^3
   e x^{n+m} f x^{n-m}
2 a 1   b 7
   c 1   d 6
3 a $10x^7$ b $4x^4$

Page 47

1 a 552   b 3172
   c 1953   d 2652
   e 6321   f 6890

Page 48

1 a 26   b 16
   c 38   d 44
   e 28   f 49

Page 49

1 a 35   b £11.34
2 a i 2 ii 3 iii 1
   b i 2.3 ii 6.1 iii 15.9
   c i 3.45 ii 16.09 iii 7.63
   d i 4.974 ii 6.216 iii 0.008

Page 50

1 a i 57.8 ii 31.1 iii 36.5 iv 3.6 v 5.83
   vi 13.7
   b i 14 ii 14.1 iii 10.8 iv 2.89 v 2.51
   vi 4.65
   c i 106.03 ii 185.32 iii 58.48 iv 1.32
   v 0.091 vi 39.76

Page 51

1 a i $\frac{19}{28}$ ii $1\frac{5}{18}$ iii $6\frac{1}{15}$
   b i $\frac{13}{30}$ ii $\frac{2}{9}$ iii $\frac{7}{12}$
   c i $\frac{1}{6}$ ii $\frac{5}{14}$ iii $3\frac{17}{20}$
   d i $\frac{7}{10}$ ii $1\frac{3}{4}$ iii $1\frac{3}{5}$
2 40

Page 52

1 a i −15 ii +24 iii −35
   b i −4 ii −2 iii +3
2 a i 2 ii 3 iii 4
   b i 4 ii 0.8 iii 60
3 a 700   b 5
   c 10   d 32
4 a 100   b 350

Page 53

1 a i 1:3 ii 5:6 iii 3:5
   b i 1:6 ii 8:1 iii 2:5
2 a £100 and £400   b 50 g and 250 g
   c £150 and £250   d 80 kg and 160 kg
3 a 56   b 30

Page 54

1 a 37.5 mph   b 52.5 km
2 a 31.25 kg   b 160
3 a Travel-size, 1.44 g/p compared to 1.38 g/p
   b 95 out of 120 is 79.2% compared to 62 out of
   80, which is 77.5%.

Page 55

1 a i 0.3 ii 0.88   b i $\frac{9}{10}$ ii $\frac{8}{25}$
   c i 85% ii 15%   d i $\frac{4}{5}$ ii $\frac{2}{25}$
   e i 0.625 ii 0.35   f i 36% ii 5%
2 a i 0.8 ii 0.07 iii 0.22
   b i 1.05 ii 1.12 iii 1.032
   c i 0.92 ii 0.85 iii 0.96

Page 56

1 a i £10.50 **ii** 19.2 kg
b i £7.20 **ii** £52.80
2 a £168 **b** 66.24 kg
3 a i 740 **ii** 5% **b** 20%
4 1.05 × 0.94 = 0.94 × 1.05

Page 59

1 22 cm, 34 cm, 22 cm
2 15–18 cm², 12–15 cm²
3 i 15 m² **ii** 16 cm²

Page 60

1 a 35 cm² **b** 12.5 m²
2 a 46 cm² **b** 26 cm²

Page 61

1 a 20 cm² **b** 35 m²
2 a $37\frac{1}{2}$ cm² **b** 34 cm²

Page 62

1 2, 1, 4, 2, 1, 1
2 4, 6, 8, 2, 4, 2
3 9, 4, 4

Page 63

1 a 80° **b** 240°
2 a 38° **b** 116°

Page 64

1 a = 75°, b = 72°, c = 36°, d = 60°, e = 22°
2 Because it forms two triangles and the angles in each triangle add up to 180°,
2 × 180° = 360°.
3 f = 90°, g = 90°, h = 117°, i = 63°, j = 109°, k = 71°
4 a = 80°, b = 90°, c = 130°, d = 130°, e = 50°

Page 65

1 a 135° **b** 140°
2 a 45° **b** 40°
3 They always add up to 180°.

Page 66

1 a d **b** e
c b **d** e
e c **f** h
g c **h** e

Page 67

1 a square, rhombus, kite
b square, rectangle, parallelogram, rhombus
c square, rhombus
d square, rectangle, parallelogram, rhombus
e square, rectangle, parallelogram, rhombus
f trapezium
g kite
2 a Yes, a square has all the properties of a rectangle.
b Yes, a rhombus has all the properties of a parallelogram.
c No, a kite does not have all sides equal.
d Yes, a square has all the properties of a rhombus.

Page 68

1 220°
2 a 270° (west) **b** 225° (south-west)
3 north-west
4 135° (south-east)

Page 69

1 a 44.0 cm **b** 37.7 cm
c 8π cm
2 a 50.3 cm² **b** 706.9 cm²
c 9π cm²

Page 70

1 a –2 °C **b** 26 mph
c 25 g
2 a i 6–7 m **ii** 8–10 m
b 1.2–1.5 kg

Page 71

1 a 150 cm **b** 1 m by 2 m
c £2.50
2 Square-based pyramid and cube.
3 A and C

Page 72

1 a 2 cm by 3 cm by 4 cm **b** 24 cm³
2 a **b** **c**

Page 73

1 a Yes   b No

  c Yes

2 Any different tessellation using at least 6 rectangles.

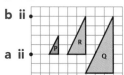

Page 74

1 a $\binom{5}{0}$ b $\binom{3}{4}$

  c $\binom{5}{-4}$ d $\binom{0}{-4}$

2 a y-axis   b $x = 3$

Page 75

1 a 90°, anticlockwise about (0, 0)

  b half-turn about (–1, 0)

2 a i 2   b i $\frac{2}{3}$

  b ii

a ii

  b iii sf $\frac{1}{3}$ and centre O

Page 76

1 a   b

c

2

3 cm 3 cm

Page 77

1 Self checking.

2

•Edinburgh

•London

Page 78

1 a 1.2 m   b 3.5 kg

  c 2.3 l   d 4.5 cm

2 a 55 lbs   b 36 l

  c 90 cm   d 75 miles

Page 79

1 a 2 m³   b 3 m²

  c 5 000 000 cm³   d 400 mm²

2 a 52 cm²   b 24 cm³

Page 80

1 5 cm

2 2.75 cm

3 a i 24 cm² ii 240 cm³

  b 42 cm³

4 160π cm³

Page 81

1 3.9 m

Page 84

1 a $r - p$ b $7 + x$

  c ab d $\frac{t}{2}$

  e $n + m$

2 a $8x$ b $9x + 5$

  c $9w - 6k$ d $2x^2 - 2z$

3 a $12t$ b $12n^3$

  c $30m^2n$ d $-12x^3y^4$

4 a Any that work, such as $2x + 3x$

  b Any that work such as $5a + 6b - 2a - 4b$

  c Any that work such as $2xy \times 3y$

Page 85

1 a $3x + 15$ b $5y - 10$

  c $6x + 3y$ d $n^2 - 7n$

  e $10m^2 + 15m$ f $6p^3 - 9p^2q$

2 a $6x + 6$ b $3m + 22$

  c $4n^2 - 5n$ d $3x + 27$

  e $4x^2 - 7xy$ f $10x + 38y$

3 a $5(n + 2m)$ b $3x(2x - 3)$

  c $m(5n + 6)$ d $4xy(x + 3y)$

  e $2x(y + 3x)$ f $2ab(a - 4 + 3b)$

4 a $15x - 12 = 3(5x - 4)$ b $5x + 10 = 5(x + 2)$

  c $6x - 4 = 2(3x - 2)$ d $-3x + 18 = 3(6 - x)$

  e $15x - 12 = 3(5x - 4)$ f $10x + 34 = 2(5x + 17)$

Page 86

1 a 3   b −1
 c 12   d $\frac{1}{2}$
 e 13   f 2
 g 13   h 28
 i $\frac{1}{2}$

Page 87

1 a 9   b 6
 c −2   d 4
 e $2\frac{1}{2}$ f $-1\frac{1}{2}$
2 a 7   b 2
 c 3   d −1
 e $\frac{1}{5}$ f $-2\frac{1}{2}$
3 a 9   b −3
 c 5   d $6\frac{1}{2}$
4 a 20   b 14
 c 5   d 5

Page 88

1 a 4   b 6
2 3
3 a $\frac{x}{2} + 7 = x + 6$ b 2

Page 89

1 4.8
2 2.6

Page 90

1 Formulae: **a**, **c**, **d** and **h**; expressions **f** and **g**; equations **b** and **e**
2 a −6   b 29
 c 24
3 a $x = \frac{T}{4}$ b $x = \frac{y-3}{2}$
 c $x = P - t$ d $x = 5y$
 e $x = \frac{A}{m}$ f $x = \frac{S}{2\pi}$

Page 91

1 a x < 3   b x > 1
 c x ⩾ 18   d x ⩽ −1
 e x > −10   f x ⩽ −1
2 a x > 1   b x ⩽ 3
3 a x < 2   b x ⩾ −1
 c −1, 0, 1

Page 92

1 a £45   b 150 units
2 a 20 minutes b The car was stationary.
 c B to C   d 45 km/h

Page 93

1 B(3, 0), C(0, 1), D(−2, 4), F(−2, 0),
 G(−4, −1), H(−3, −3), I(1, −3)
2 a −5, −3, −1
 b
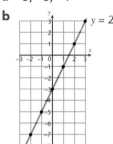

3
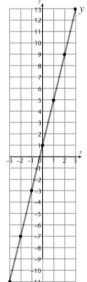

Page 94

1 a A and C

 b A and B

2 a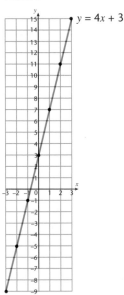

 b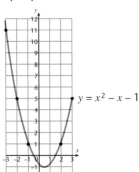

Page 95

1 a 1, −1, −1

 b

$y = x^2 - x - 1$

Page 96

1 a +1.4, −1.4

 b 4.4 and −1.4

Page 97

1 a $12345 \times 9 + 6 = 111111$,
 $123456 \times 9 + 7 = 1111111$

 b $123456 \times 8 + 6 = 987654$,
 $1234567 \times 8 + 7 = 9876543$

 c $5 \times 3 \times 37 = 555$, $6 \times 3 \times 37 = 666$

 d $66667 \times 66667 = 4444488889$,
 $666667 \times 666667 = 444444888889$

2 a i 39, 45 ii goes up in 6s

 b i 96, 192 ii doubles

 c i 76, 72 ii down in 4s

 d i 59, 75 ii up 2, 4, 6, 8 …

 e i −2, −5 ii down in 3s

Page 98

1 a i 3, 7, 11 ii 399

 b i 3, 6, 10 ii 20 100

2 a $5n + 1$

 b $8n - 5$

 c $3n + 6$

3 The first sequence increases by 5 and the second by 10 which is a multiple of 5.

Page 99

1 a 200

 b 199

 c 10 000

 d 36, 45, 55, 66, 78

 e 256, 512, 1024, 2048, 4096

 f 23, 29, 31, 37, 41

2 a 41

 b $4n + 1$

NEW GCSE
MATHS
Foundation

Exam Practice Workbook

For GCSE Maths from 2010

Edexcel + AQA + OCR

Keith Gordon

Statistical representation

G

1 Zeke did a survey of the number of passengers in some cars passing the school.

The results are shown in the table.

Number of passengers	Tally	Frequency
1	ℍℍ ℍℍ ℍℍ ℍℍ /	21
2	ℍℍ ℍℍ ///	
3		8
4	ℍℍ /	
5	//	

a Complete the frequency column. **[1 mark]**

b Complete the tally column. **[1 mark]**

c How many cars were surveyed altogether? _____ **[1 mark]**

AU d Explain why the total number of passengers in all the cars was 105.

_____

_____ **[2 marks]**

G

2 The pictogram shows the number of letters delivered to a company during five days.

⊠ represents 20 letters.

Day		Number of letters
Monday	⊠ ⊠ ⊠ ▷	65
Tuesday	⊠ ⊠ ⊠ ⊠ ⊠	
Wednesday	⊠ ⊠ ⊠ ◹	
Thursday		40
Friday		35

a Complete the 'number of letters' column. **[1 mark]**

b Complete the pictogram column. **[1 mark]**

c How many letters were delivered altogether during the week? _____ **[1 mark]**

C

AU 3 Describe the method of data collection you would use to investigate the following.

a The likelihood of an earthquake in Hawaii _____

b The fairness of a home-made spinner _____

c The probability of winning a prize in a raffle _____

d The percentage of people in a small town who would like to see a new supermarket

opened. _____ **[3 marks]**

G

FM 1 This table shows the numbers of hours ten students spent watching TV and doing homework one weekend.

Student	A	B	C	D	E	F	G	H	I	J
Hours TV	2.5	1.5	4	3.5	5	1	3	4.5	3	2.5
Hours homework	1.5	3	1	1.5	0.5	2.5	1.5	0.5	2	1.5

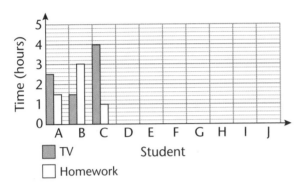

TV

Homework

a Complete the dual bar chart to illustrate the data. **[2 marks]**

b The school recommends that students spend at least $1\frac{1}{2}$ hours doing homework each weekend. How many students did at least $1\frac{1}{2}$ hours homework?

_____ **[1 mark]**

c Bernice says that all the students spent more time watching TV than doing homework. Is this true? Explain your answer.

_____ **[1 mark]**

FM 2 The line graph shows how the temperature in a greenhouse varies.

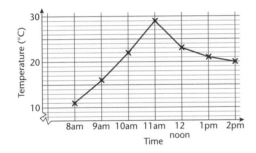

F

a What was the temperature at 9am?

_____ °C **[1 mark]**

b During which hour did the temperature increase the most?

_____ **[1 mark]**

c The gardener opened the ventilator to lower the temperature. At what time did he do this?

_____ **[1 mark]**

d What was the approximate temperature at 10.30am? _____ °C **[1 mark]**

e Can you use the graph to predict the temperature at 6pm? Explain your answer.

_____ **[1 mark]**

AU 3 The first bar chart show the stocks of replica-shirts in stock when a football club shop opens one day. The second bar chart shows how many of each shirt were left at the end of the day.

14 heritage shirts were sold that day.

How many shirts were sold altogether?

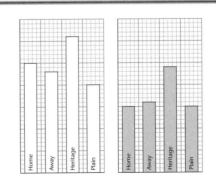

_____ **[3 marks]**

E

Statistics

F

1 The table shows the numbers of passengers in some cars.

Number of passengers	Frequency
1	23
2	15
3	5
4	4
5	3

a How many cars were in the survey?

_____ **[1 mark]**

b What is the modal number of passengers per car?

_____ **[1 mark]**

c What is the median number of passengers per car?

_____ **[1 mark]**

d Cars with two or more passengers can use a 'car pool' lane on the motorway.
What percentage of these cars could use the car pool lane?

_____ **[1 mark]**

C

2 a For the data 51, 74, 53, 74, 76, 58, 68, 51, 70 and 65 work out:

   i the mean

_____ **[2 marks]**

   ii the range

_____ **[1 mark]**

b A football team of 11 players has a mean weight of 84 kg.

   i How much do the 11 players weigh altogether?

_____ **[1 mark]**

   ii When the three substitutes are included, the 14 players have a mean weight
of 87 kg. What is the mean weight of the three substitutes?

_____ **[3 marks]**

E

PS 3 Find five numbers with a mean of 8, a mode of 9 and a range of 4.

[3 marks]

1 The bar chart shows the numbers of spoonfuls of sugar that a group of workmen take in their morning tea.

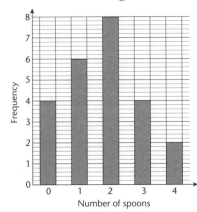

a How many workmen are there?

_____ [1 mark]

b What is the modal number of spoons?

_____ [1 mark]

c What is the median number of spoons?

_____ [1 mark]

d What is the mean number of spoons?

_____ [2 marks]

e The workmen buy a kilogram of sugar. This is enough for 400 spoons of sugar. Will they have enough sugar for five days if they each have two cups of tea a day? Explain your answer fully.

_____ [2 marks]

D

2 Ten workmates went ten-pin bowling.

a Their scores for the first game were: 87 123 121 103 93 231 145 46 65 46
Work out:

i   the modal score   _____ [1 mark]

ii the median score   _____ [1 mark]

iii the mean score.   _____ [2 marks]

b **i** Explain why the mode would not be a good average to use.

_____ [1 mark]

ii Explain why the mean would not be a good average to use.

_____ [1 mark]

c In the second game, the modal score was 105. Does this mean the players increased their overall scores? Explain your answer.

_____ [1 mark]

C

AU 3 Six numbers have a mean of 8, a mode of 7 and a median of 7.5.
A seventh number is added to the set. You are told three new pieces of information about the 7 numbers.

Fact A: The new mean is 9.
Fact B: The new mode is 7.
Fact C: The new median is 8.

a What number was added to the set?   _____ [2 marks]

b Which of the facts above are not needed to find the new value?   _____ [1 mark]

C

D

1 The table shows the number of cars per house on a housing estate of 100 houses.

Number of cars	Number of houses
0	8
1	23
2	52
3	15
4	2

Work out:

a the modal number of cars

_____ **[1 mark]**

b the median number of cars

_____ **[1 mark]**

c the mean number of cars.

_____ **[2 marks]**

C

2 The table shows the scores of 200 boys in a mathematics exam.

The frequency polygon shows the scores of 200 girls in the same examination.

Mark, x	Frequency, f
$40 < x \leq 50$	27
$50 < x \leq 60$	39
$60 < x \leq 70$	78
$70 < x \leq 80$	31
$80 < x \leq 90$	13
$90 < x \leq 100$	12

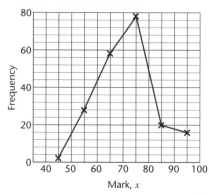

a Work out the mean mark for the boys' scores. _____ **[4 marks]**

b On the same graph as the girls frequency polygon, draw the frequency polygon for the boys' scores. **[2 marks]**

c Who did better in the test, the boys or the girls? Give reasons for your answer.

_____

_____ **[1 mark]**

C

AU 3 The histogram shows the weights of 50 members of a sports club.

Calculate the mean weight.

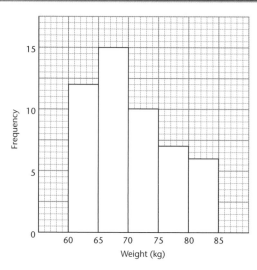

_____ **[4 marks]**

D

1 These are the weights of 20 guinea pigs, in grams, rounded to the nearest 10 grams.

130   90   220   210   190   130   160   110   230   90

80   120   130   240   180   150   70   220   240   130

a Using the key 1 | 3 to represent 130 grams, put the data into the stem-and-leaf diagram.

0 |

1 |

2 | **[2 marks]**

Key 1 | 3 represents 130 grams

b Using the stem-and-leaf diagram, or otherwise, write down:

i the modal weight   _____ grams   **[1 mark]**

ii the median weight   _____ grams   **[1 mark]**

iii the range of the weights.   _____ grams   **[1 mark]**

C

2 A teacher recorded how many times her students were late during a term.

The stem-and-leaf diagram shows the data.

12 students were **never late**.

0 | 2   3   4   4   5   6   7

1 | 3   5   8   9   9

2 | 0   1   4   5

3 | 2

5 | 1

Key 1 | 7 represents 17 times late

a How many students there are in the form altogether?   _____   **[1 mark]**

b Work out the mean number of times late for the whole
form.   _____   **[3 marks]**

AU 3 The stem-and-leaf diagram shows the number of matches in 10 boxes.

3 | 7   8   9

4 | 0   1   2   2   2   4   5

Key 3 | 8 represents 38 matches

The contents of another box are added to the stem-and-leaf diagram.

D

a Will the mode change?

☐ Yes   ☐ No   ☐ Cannot tell   Justify your choice of answer.   **[1 mark]**

b Will the median change?

☐ Yes   ☐ No   ☐ Cannot tell   Justify your choice of answer.   **[1 mark]**

c Will the mean change?

☐ Yes   ☐ No   ☐ Cannot tell   Justify your choice of answer.   **[1 mark]**

Statistics

G

1 a State whether each of the following events is *impossible, very unlikely, unlikely, evens, likely, very likely* or *certain*.

   i   you walking on the moon tomorrow _____ [1 mark]

   ii getting a six when a regular dice is thrown _____ [1 mark]

   iii tossing a coin and scoring a head _____ [1 mark]

   iv someone in the class going abroad for their holidays _____ [1 mark]

b On the probability scale, put a numbered arrow to show approximately the probability of each of the following outcomes of events happening.

```
 ┌─────────────────────────────┬─────────────────────────────┐
 0                             1/2                            1
```

   i   The next car you see driving down the road will only have the driver inside.   **[1 mark]**

   ii Someone in the class had porridge for breakfast.   **[1 mark]**

   iii Picking a red card from a well shuffled pack of cards.   **[1 mark]**

   iv Throwing a number less than seven with a regular dice.   **[1 mark]**

E

2 A bag contains 20 coloured balls. Twelve are red, five are blue and the rest are white. A ball is taken from the bag at random.

a What is the probability that the ball is:

   i   red _____ [1 mark]

   ii pink _____ [1 mark]

   iii blue or white? _____ [1 mark]

b Some more white balls are added to the bag so that the probability of getting a red ball is now $\frac{1}{2}$. How many white balls were added?

_____ [1 mark]

C

AU 3 The ratio of red counters to blue counters in a bag is 2 : 3.

a What is the probability of taking a red counter at random from the bag?

_____ [1 mark]

b It is known that there are 10 counters in the bag. How many blue counters must be added so that the probability of taking a red counter is now $\frac{1}{4}$?

✓ _____ [2 marks]

C

AU 4 The ratio of white balls to black balls in a bag is 4 : 5.

Doris says that the probability of picking a white ball is $\frac{4}{5}$.

Explain why Doris is wrong.

_____ [1 mark]

Using probability

1 The probability that a milkman delivers the wrong sort of milk to a house is $\frac{3}{50}$.

a What is the probability that he delivers the correct sort of milk to a house?

_____ **[1 mark]**

b He delivers milk to 500 houses a day. Estimate the number of houses that get the wrong milk.

_____ **[1 mark]**

2 There are 900 squares on a 'Treasure map' at the school Summer Fayre.

One of the squares contains the treasure.

The Rogers decide to buy some squares.

Mr Rogers buys five squares, Mrs Rogers buys ten squares and their two children, Amy and Ben, buy two squares each.

a Which member of the family has the best chance of winning?

Explain your answer.

_____ **[1 mark]**

b What is the probability that Mr Rogers wins the treasure?

_____ **[1 mark]**

c What is the probability that one of the children wins the treasure?

_____ **[1 mark]**

d If the family put all their squares together, what is the probability that they will win the treasure?

_____ **[1 mark]**

3 John makes a dice and weights one side with a piece of sticky gum. He throws it 120 times. The table shows the results.

Score	1	2	3	4	5	6
Frequency	18	7	22	21	35	17
Relative frequency						

a Fill in the relative frequency row. Give your answers to 2 decimal places. **[2 marks]**

b Which side did John stick the gum on? Explain how you can tell.

_____ **[1 mark]**

PS 4 Three bags contain white and black balls.

Bag A contains three white and three black balls.

Bag B contains three white and four black balls.

Bag C contains four white and five black balls.

Two bags are to be combined together.

Which two bags should be combined to give the greatest chance of picking a white ball?

_____ **[3 marks]**

Statistics

E

1 Pete's Café has a breakfast deal.

Three-item breakfast! Only £1		
Choose one of:	**Choose one of:**	**Choose one of:**
sausage or bacon	egg or hash browns	beans or toast

a There are eight possible 'three-item breakfast' combinations, for example, **sausage**, **egg**, **beans** or **sausage**, **egg**, **toast**.

List all the other possible combinations.

_____   _____

_____   _____

_____   _____

_____   _____ **[2 marks]**

b Fred tells his friend, 'I'll have what you are having.'

What is the probability that Fred gets bacon and eggs with his breakfast?

_____ **[1 mark]**

D

2 The sample space diagram shows the outcomes when a coin and a regular dice are thrown at the same time.

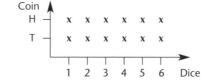

a How many possible outcomes are there when a coin and a dice are thrown together?

_____ **[1 mark]**

b When a coin and a dice are thrown together what is the probability that:

i the coin shows tails and the dice shows an even number

_____ **[1 mark]**

ii the coin shows heads and the dice shows a square number?

_____ **[1 mark]**

c Alicia and Zeek play a game with the coin and dice. If the coin lands on a head the score on the dice is doubled. If the coin lands on a tail 1 is subtracted from the score on the dice.

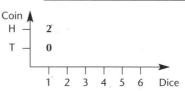

i Complete the sample space diagram to show all possible scores. **[2 marks]**

ii What is the probability of an odd score?

_____ **[1 mark]**

C

(PS 3) Katy needs to throw a score of 5 to win a game. She can choose to throw one dice or two dice. Which gives the best chance of getting a score of 5, throwing one dice or throwing two dice?

_____ **[4 marks]**

Expectation and two-way tables

1 A bag contains 30 balls that are either red or white. The ratio of red balls to white balls is 2:3.

   a Zoe says that the probability of picking a red ball at random from the bag is $\frac{2}{3}$.

      Explain why Zoe is wrong.

      _____

      _____ **[1 mark]**

   b How many red balls are there in the bag?

      _____ **[1 mark]**

   c A ball is taken from the bag at random, its colour noted and then it is replaced.

      This is done 200 times. How many of the balls would you expect to be red?

      _____ **[1 mark]**

---

2 The two-way table shows the numbers of male and female teachers in four school departments.

	Male	Female
Mathematics	7	5
Science	11	7
RE	1	3
PE	3	3

   a How many male teachers are there altogether? _____ **[1 mark]**

   b Which subject has equal numbers of male and female teachers?

      _____ **[1 mark]**

   c Nuna says that science is a more popular subject than mathematics for female teachers.

      Explain why Nuna is wrong.

      _____

      _____ **[1 mark]**

   d What is the probability that a teacher chosen at random from the table will be a female teacher of mathematics or science?

      _____

      _____ **[1 mark]**

PS 3 Here is some information about an exercise class of 25 people.

- There are five more women than men
- A third of the women are 65 or over.
- Four times as many men are under 65 than 65 or over.

Use the information to fill in the two-way table.

	Men	Women	Total
65 or over	2	5	
Under 65	8	10	
Total	10	15	

[3 marks]

D

1   The table shows the results of a survey of 60 students, to find out what they do for lunch.

Lunch arrangement	Frequency	Angle
Use canteen	22	
Have sandwiches	18	
Go home	12	
Go to shopping centre	8	

Tom is going to draw a pie chart to show the data.

a   Complete the column for the angle for each sector.   **[2 marks]**

b   Draw a fully labelled pie chart below.

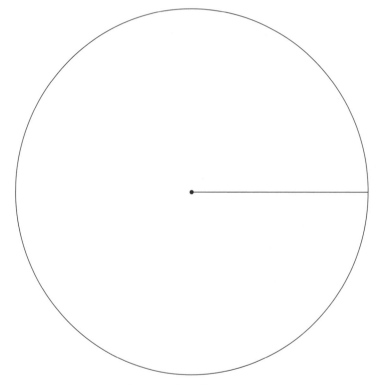

[2 marks]

c   There are 1200 students in the school.

Estimate how many of them go to the shopping centre at lunchtime.

_____ **[1 mark]**

C

AU 2   Matt is drawing a pie chart. One sector has a frequency of 15.

a   Matt calculates that the angle will be 80°.
Explain why Matt must have made a mistake.

_____ **[1 mark]**

b   Which of the following could be the angle of a sector with a frequency of 15?

60   75   100   120   144   **[1 mark]**

Scatter diagrams

FM 1 A delivery driver records the distances and times for deliveries.

The table shows the results.

Delivery	A	B	C	D	E	F	G	H	I	J
Distance (km)	12	16	20	8	18	25	9	15	20	14
Time (minutes)	30	42	55	15	40	60	45	32	20	35

a Plot the values on the scatter diagram.

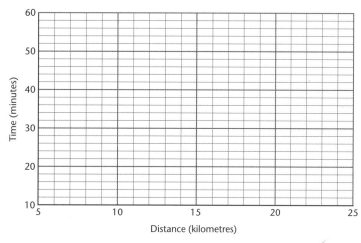

[2 marks]

b i During one of the deliveries the driver was stuck in a traffic jam. Which delivery was this? _____ [1 mark]

ii One of the deliveries was done very early in the morning when there was no traffic. Which delivery was this? _____ [1 mark]

c Ignoring the two values in **b**, draw a line of best fit through the rest of the data. [1 mark]

d Under normal conditions, how long would you expect a delivery of 22 kilometres to take? _____ [1 mark]

AU 2 Match each of these scatter diagrams to one of the statements below.

A 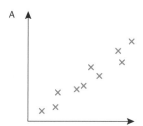   B 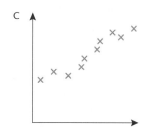   C   D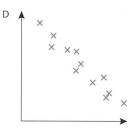

W: How much is spent on school lunch against distance lived from the school.

_____

X: Percentage value of the initial cost of a car against the age of the car.

_____

Y: Marks in a mathematics exam against marks in a science exam.

_____

Z: Cost of a taxi journey against distance

_____ [2 marks]

1 Danny wanted to investigate the hypothesis:

Girls spend more time on mathematics coursework than boys do.

a Design a two-way table that will help Danny collect data.

[2 marks]

b Danny collected data from 30 boys and 10 girls.

He found that, on average, the boys spent 10 hours and the girls spent 11 hours on their mathematics coursework.

Does this prove the hypothesis? Give reasons for your answer.

_____ **[1 mark]**

2 Daz does a survey about school lunches.

a This is one of the questions in his survey.

Burgers and chips are popular but not healthy.

Don't you agree? ☐ Yes   ☐ No   Give two criticisms of this question

Criticism 1 _____

Criticism 2 _____ **[2 marks]**

b This is another of his questions.

How much do you spend, on average, on school lunch? _____

Design a response section for this question

[2 marks]

(**FM 3**) A farmer thinks that the amount of rainfall is increasing each year.

He records the amount of rain that falls each month for a year.

The table shows the results (in mm).

	Jan	Feb	Mar	Apr	May	Jun	Jul	Aug	Sep	Oct	Nov	Dec
2009	39	45	32	38	56	22	18	21	34	39	45	43

On the internet he finds out that the mean rainfall for 2008 was 35 mm, with a range of 40.

Investigate the hypothesis:

'The amount of rainfall is increasing each year.'

[5 marks]

Statistics record sheet

Name _____ **Marks** _____

Form _____ **Percentage** _____

Date _____ **Grade** _____

Page number	Question number	Topic	Mark	Comments
112	1	Statistical representation	/5	
	2	Statistical representation	/3	
	3	Statistical representation	/3	
113	1	Statistical representation	/4	
	2	Statistical representation	/5	
	3	Statistical representation	/3	
114	1	Averages	/4	
	2	Averages	/7	
	3	Averages	/3	
115	1	Averages and range	/7	
	2	Averages and range	/7	
	3	Averages and range	/3	
116	1	Arranging data	/4	
	2	Arranging data	/7	
	3	Arranging data	/4	
117	1	Arranging data	/5	
	2	Arranging data	/4	
	3	Arranging data	/3	
118	1	Probability	/8	
	2	Probability	/4	
	3	Probability	/3	
	4	Probability	/1	
119	1	Using probability	/2	
	2	Using probability	/4	
	3	Using probability	/3	
	4	Using probability	/3	
120	1	Combined events	/3	
	2	Combined events	/6	
	3	Combined events	/4	
121	1	Expectation and two-way tables	/3	
	2	Expectation and two-way tables	/4	
	3	Expectation and two-way tables	/3	
122	1	Pie charts	/5	
	2	Pie charts	/2	
123	1	Scatter diagrams	/6	
	2	Scatter diagrams	/2	
124	1	Surveys	/3	
	2	Surveys	/4	
	3	Surveys	/5	
Total			**/159**	

Successes

1 _____

2 _____

3 _____

Areas for improvement

1 _____

2 _____

3 _____

G

1 a Write down the answer to each calculation.

    i 8×9 _____ **[1 mark]**

    ii 40×7 _____ **[1 mark]**

    iii $240 \div 3$ _____ **[1 mark]**

AU b George has four cards with a number written on each of them.

    i He uses the cards to make a multiplication statement.

 $\times$ $=$

       What is the multiplication statement?   _____ **[1 mark]**

    ii He uses the cards to make a division statement.

  $\div$ $=$

       What is the division statement?   _____ **[1 mark]**

D

PS 2 Two students work out the following calculation:

$2 + 4^2 \div 8$

Sammi gets an answer of 2.25. Ross gets an answer of 4.5.

Both of these answers are wrong.

a What is the answer to $2 + 4^2 \div 8$?   _____ **[1 mark]**

b Put brackets in the following calculations to make them true.

    i $2 + 4^2 \div 8 = 2.25$ **[1 mark]**

    ii $2 + 4^2 \div 8 = 4.5$ **[1 mark]**

F

PS 3 The diagram shows a normal dartboard.

a 20 and 1 are adjacent numbers, and $20 + 1 = 21$.

    i Find the highest total made by two adjacent numbers.   _____ **[1 mark]**

    ii Find the smallest possible total made by two adjacent numbers.   _____ **[1 mark]**

b 20 and 3 are opposite numbers, and $20 + 3 = 23$.

    i Find the highest total made by two opposite numbers.   _____ **[1 mark]**

    ii Find the smallest possible total made by two opposite numbers.   _____ **[1 mark]**

1 Work these out.

a
```
  2 5 7 6
+ 1 0 8 3
─────────
```

b
```
    1 2 9
+ 6 7 3 5
─────────
```

c $78 + 2054 - 362$

[1 mark each]

G

2 a When England won the world cup at Wembley in 1964 the attendance was given as 96 924.

   i Write 96 924 in words.

      _____ [1 mark]

   ii Round 96 924 to the nearest 100. _____ [1 mark]

   iii Round 96 924 to the nearest 1000. _____ [1 mark]

b When Italy won the world cup in Berlin in 2006, the attendance was given as 69 000 rounded to the nearest hundred.

   i What is the value of the digit 9 in 69 000? _____ [1 mark]

   ii What is the smallest value the attendance could have been? _____ [1 mark]

   iii What is the largest value the attendance could have been? _____ [1 mark]

   iv About one third of the spectators were Italian.

       Approximately how many spectators were Italian? _____ [1 mark]

G

3 Work these out.

a
```
  3 0 7 6
- 2 1 7 8
─────────
```

b
```
  6 9 0 3
- 3 7 2 5
─────────
```

c $86 + 1623 - 484$

[1 mark each]

F

4 Farmer Bill has 1728 sheep on his farm. Farmer Jill has 589 sheep on her farm.

a How many more sheep does farmer Bill have than farmer Jill?

      _____ [1 mark]

b Farmer Jill sells all of her sheep to farmer Bill. How many does he have now?

      _____ [1 mark]

F

PS 5 This year I am twice as old as my daughter.

Fourteen years ago I was four times as old as she was then.

How old am I now?

      _____ [3 marks]

C

1 Work these out.

   a 7 6   **b**   **c** 54×7 **d** $384 \div 8$

       $\times$ 4 $6\,)\overline{156}$

      ‾‾‾‾   **[1 mark each]**

2 a Mary buys four cans of cola at 68p per can.

    i How much do the four cans cost altogether?   _____ **[1 mark]**

    ii She pays with a £5 note. How much change does she get?

                                                                      _____ **[1 mark]**

  b In the school hall, there are 24 chairs in each row.

      There are 30 rows of chairs.

      How many chairs are there in total?   _____ **[2 marks]**

  c Year 10 has 196 students in seven forms.

      Each form has the same number of students.

      How many students are there in each form?   _____ **[2 marks]**

3 Show the calculation you need to do to work out each answer.

Then calculate the answer.

  a How much change do I get from £20 if I spend £12.85?

      _____

      _____ **[1 mark]**

  b I buy three ties at a total cost of £14.55. What is the price of each tie?

      _____

      _____ **[2 marks]**

  c Cartons of eggs contain 12 eggs. How many eggs will there be in nine cartons?

      _____

      _____ **[2 marks]**

AU 4 Here are three decimals.

    A 3.65   B 3.725   C 0.3627

  a Give a mathematical property that A and B have in common.

                                                       _____ **[1 mark]**

  b Give a mathematical property that B and C have in common.

                                                       _____ **[1 mark]**

  c Which is larger, $3.65 \div 10$ or 0.3627?

                                                       _____ **[1 mark]**

1 a Which two of these fractions are equivalent to $\frac{3}{4}$?

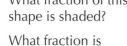

_____ and _____ **[1 mark]**

b Shade $\frac{3}{4}$ of this shape.

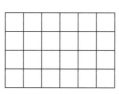

[1 mark]

c i What fraction of this shape is shaded?

_____ **[1 mark]**

ii What fraction is not shaded?

_____ **[1 mark]**

2 a i Shade $\frac{2}{9}$ of this shape.

[1 mark]

ii Shade $\frac{1}{3}$ of this shape.

[1 mark]

b Use your answer to part (a) to write down the answer to:

$\frac{2}{9} + \frac{1}{3} =$ _____ **[1 mark]**

3 The ratio of grey squares to white squares in this shape is 2:3.

a Zoe says, 'That means the grey squares must be $\frac{2}{3}$ of the shape.'

Explain why Zoe is wrong.

_____ **[1 mark]**

b Write down **two** other fractions that are equivalent to $\frac{2}{5}$.

_____ and _____ **[1 mark]**

AU 4 Here are three fractions.

$\frac{3}{2}$ $\frac{2}{3}$ $\frac{2}{8}$

Give a reason why each of them could be the odd one out.

_____

_____

_____

[3 marks]

G

1 a Shade squares in each of these diagrams so that $\frac{1}{4}$ of each diagram is shaded.

    i 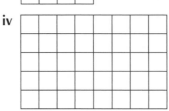   **ii**

    iii   **iv**

[2 marks]

b Fill in the boxes to make the following fractions equivalent.

    i $\dfrac{3}{7} = \dfrac{\square}{28}$ **ii** $\dfrac{5}{8} = \dfrac{20}{\square}$

    iii $\dfrac{15}{18} = \dfrac{5}{\square}$ **iv** $\dfrac{\square}{3} = \dfrac{16}{24}$ [1 mark each]

c Cancel the following fractions, giving each answer in its simplest form.

    i $\frac{16}{28} = $ _____   **ii** $\frac{9}{15} = $ _____

    iii $\frac{12}{30} = $ _____   **iv** $\frac{21}{28} = $ _____   [1 mark each]

d Put the following fractions in order, with the smallest first.

    $\frac{7}{10}$ $\frac{4}{5}$ $\frac{13}{20}$ _____ [1 mark]

F

2 a Change the following top-heavy fractions into mixed numbers.

    i $\frac{9}{5} = $ _____   **ii** $\frac{17}{7} = $ _____

    iii $\frac{21}{8} = $ _____   **iv** $\frac{31}{4} = $ _____   [1 mark each]

b Change the following mixed numbers into top-heavy fractions.

    i $1\frac{6}{11} = $ _____   **ii** $1\frac{3}{8} = $ _____

    iii $2\frac{1}{3} = $ _____   **iv** $4\frac{3}{5} = $ _____   [1 mark each]

C

PS 3 What fraction of this diagram is shaded?

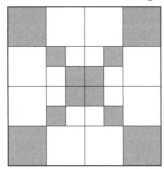

_____ [3 marks]

E

FM 1 **a** Fill in the boxes to make the following fractions equivalent.

i $\frac{3}{10} = \frac{\square}{20}$ ii $\frac{5}{8} = \frac{\square}{16}$ iii $\frac{1}{3} = \frac{\square}{6}$ iv $\frac{3}{4} = \frac{\square}{12}$ **[1 mark each]**

b Use the answers to part **a** to work these out.

Cancel the answer to its simplest form.

i $\frac{1}{20} + \frac{3}{10}$ _____ **[1 mark]**

ii $\frac{3}{16} + \frac{5}{8}$ _____ **[1 mark]**

iii $\frac{1}{6} + \frac{1}{3}$ _____ **[1 mark]**

iv $\frac{1}{12} + \frac{3}{4}$ _____ **[1 mark]**

c Use the answers to part **a** to work these out.

Cancel the answer to its simplest form.

i $\frac{17}{20} - \frac{3}{10}$ _____ **[1 mark]**

ii $\frac{15}{16} - \frac{5}{8}$ _____ **[1 mark]**

iii $\frac{5}{6} - \frac{1}{3}$ _____ **[1 mark]**

iv $\frac{11}{12} - \frac{3}{4}$ _____ **[1 mark]**

2 **a** On MacDonald's Farm there are a total of 245 hectares.

$\frac{2}{5}$ of the land is planted for crops, the rest is used for animals.

i What fraction of the land is used for animals?   _____ **[1 mark]**

ii How many hectares are used for crops?   _____ **[2 marks]**

b MacDonald has 120 sheep.

$\frac{3}{4}$ of the sheep each gives birth to two lambs, the rest give

birth to one lamb.

How many lambs are born altogether?   _____ **[2 marks]**

c MacDonald has 220 cows. He sells $\frac{1}{4}$ of them.

How many cows has he got left after that?   _____ **[2 marks]**

F

PS 3 Mary eats $\frac{1}{5}$ of a cake.

John then eats $\frac{1}{4}$ of what is left.

Raz then eats $\frac{1}{3}$ of what is left.

Zavi then eats $\frac{1}{2}$ of what is left.

Art then gets the final piece.

Who got the most cake?

_____ **[3 marks]**

D

E

1 a Work these out, giving each answer in its simplest form.

    i $\frac{2}{3} \times \frac{6}{11}$ _____ **[1 mark]**

    ii $\frac{3}{8} \times \frac{4}{9}$ _____ **[1 mark]**

    iii $\frac{5}{6} \times \frac{3}{20}$ _____ **[1 mark]**

    iv $\frac{9}{10} \times \frac{5}{6}$ _____ **[1 mark]**

b Work out each of these and give the answer as a mixed number in its simplest form.

    i $4 \times \frac{3}{8}$ _____ **[1 mark]**

    ii $5 \times \frac{3}{10}$ _____ **[1 mark]**

    iii $6 \times \frac{1}{3}$ _____ **[1 mark]**

    iv $8 \times \frac{3}{4}$ _____ **[1 mark]**

E

2 a In a school there are 1500 students. 250 of the students are in Year 7.

    What fraction of the students are in Year 7?   _____ **[2 marks]**

b There are 120 girls in Year 7. 40 of the girls are left-handed.

    What fraction of the girls are left-handed?   _____ **[2 marks]**

E

3 a Frank earns £18 000 per year. He pays $\frac{3}{20}$ of his pay in tax.

    How much tax does he pay?   _____ **[2 marks]**

b Packets of washing powder normally contain 1.2 kg.

    A special offer pack contains $\frac{1}{6}$ more than a normal pack.

    How much does the special offer pack contain?   _____ **[2 marks]**

C

4 A packet of washing powder normally contains 500 g and costs £1.20.

There are two special offers on the washing powder:

Offer A: $\frac{1}{5}$ extra for the same price

Offer B: Same weight for $\frac{4}{5}$ of the original price.

Which offer is the best value?

                                                                         _____ **[3 marks]**

Rational numbers

1 a Write each fraction as a decimal. Give the answer as a terminating decimal or a recurring decimal as appropriate.

D

    i $\frac{7}{40}$ _____ [1 mark]

    ii $\frac{11}{15}$ _____ [1 mark]

    iii $\frac{5}{6}$ _____ [1 mark]

    iv $\frac{9}{50}$ _____ [1 mark]

b $\frac{1}{9} = 0.1111\ldots$ $\frac{2}{9} = 0.2222\ldots$

  Use this information to write down:

    i $\frac{4}{9}$ _____ [1 mark]

    ii $\frac{5}{9}$ _____ [1 mark]

c $\frac{1}{11} = 0.0909\ldots$ $\frac{2}{11} = 0.1818\ldots$

  Use this information to write down:

    i $\frac{3}{11}$ _____ [1 mark]

    ii $\frac{6}{11}$ _____ [1 mark]

---

2 a Put the following numbers in order, starting with smallest.

  $\frac{11}{20}$ $\frac{6}{11}$ $\frac{14}{25}$ $\frac{5}{9}$ _____ [2 marks]

C

b Put the following numbers in order, starting with the largest.

  $\frac{1}{4}$ $\frac{2}{9}$ $\frac{3}{11}$ $\frac{6}{25}$ _____ [2 marks]

PS 3   **a**   Work these out.

C

    i $1 \div 1.25$ _____

    ii $1 \div 2.5$ _____

    iii $1 \div 5$ _____

    iv $1 \div 10$ _____

    [2 marks]

b The sequence 1.25, 2.5, 5, 10, … is formed by doubling each term to find the next term.

  Using your answers to part **a**, explain how you can find the answer to $1 \div 40$ without using your calculator.

  _____ [1 mark]

Number

Negative numbers

AQA 1/2   EDEXCEL 1/2/3   OCR 2/3

F

1 These maps show the maximum and minimum temperatures in five towns during a six-month period.

Minimum
- Aberdeen −8°C
- Edinburgh −6°C
- Leeds −1°C
- Bristol 1°C • London 0°C

Maximum
- Aberdeen 22°C
- Edinburgh 26°C
- Leeds 28°C
- Bristol 31°C • London 33°C

a Which town had the lowest minimum temperature? _____ **[1 mark]**

b What is the difference between the lowest and highest **minimum** temperatures?

_____ **[1 mark]**

c Which two towns had a difference of 30 degrees between the maximum and minimum temperatures?

_____ **[1 mark]**

d Which town had the greatest difference between the maximum and minimum temperatures?

_____ **[1 mark]**

F

2 a The number line has the value −2.2 marked on it.

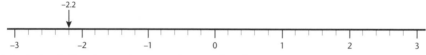

−2.2

−3   −2   −1   0   1   2   3

Mark the following values on the number line.

i −1.4   **ii** 1.7   **iii** −0.3   **[2 marks]**

b Fill in the missing values on this number line.

−1.9   −1.8   −1.7

[1 mark]

c What number is halfway between:

i −4 and 8   **ii** −11 and −8?

_____   _____ **[1 mark each]**

F

AU 3 What numbers are being described?

More than −7

Less than −1

Divides exactly by 3

_____ **[2 marks]**

1 Look at the following number cards.

+8   +6   +3   0   ⁻5   ⁻6   ⁻7

a What is the total of the numbers on all the cards?

_____ **[1 mark]**

b Which two cards will make this calculation true?

 + = **0**

_____ and _____ **[1 mark]**

c i Which card would make the answer to this calculation as small as possible?

 +4 − =

_____ **[1 mark]**

ii What is the smallest possible answer? _____ **[1 mark]**

d i Which card would make the answer to this calculation as large as possible?

 +4 − =

_____ **[1 mark]**

ii What is the largest possible answer? _____ **[1 mark]**

2 In this magic square the numbers in each row, column and diagonal add up to the same total.

Fill in the missing numbers.

3		
−3		
−6	7	−7

[2 marks]

AU 3 **a** Write two **positive** numbers in the boxes to make the equation true.

☐ − ☐ = −5 **[1 mark]**

b Write two **negative** numbers in the boxes to make the equation true.

☐ − ☐ = −5 **[1 mark]**

1 Here are six number cards.

6 **7** **10** **11** **12** **13**

a Which **two** of the numbers are multiples of 3? _____ **[1 mark]**

b Which **two** of the numbers are factors of 30? _____ **[1 mark]**

c In this magic square the numbers in each row, column and diagonal add up to the same total. Use numbers from the cards above to complete the magic square.

		8
5	9	

[2 marks]

2 a Write down the factors of each number.

   i 33 _____ **[1 mark]**

   ii 18 _____ **[1 mark]**

b From this list of numbers:

   84, 85, 86, 88, 89, 90

   write down:

   i a multiple of 3 _____ **[1 mark]**

   ii a multiple of 5. _____ **[1 mark]**

c Counter A counts a beat every 3 seconds.

   Counter B counts a beat every 4 seconds

   Counter C counts a beat every 5 seconds.

   They all start at the same time.

   After how many seconds will all three counters next count a beat at the same time? _____ **[1 mark]**

(AU 3) A teacher is describing a number to her class.

> It is a multiple of 3

> It is odd

> It is a factor of 24

What number is being described?

_____ **[1 mark]**

1 Here are seven number cards.

  6   **9**   **10**   **11**   **13**   **15**   **16**

    a   Which **two** of the numbers are prime numbers?   _____ **[1 mark]**

    b   Jen says that all prime numbers are odd.

         Give an example to show that Jen is wrong.   _____ **[1 mark]**

    c   Which **two** of the numbers are square numbers?   _____ **[1 mark]**

    d   Ken says that all square numbers end in the digits 1, 4, 6 or 9.

         Give an example to show that Ken is wrong.

         _____ **[1 mark]**

    e   Which **two** of the cards will give a result that is a prime number in the calculation below?

    6   +   =   prime

                                                       _____ **[2 marks]**

2 P is a prime number. S is a square number, Q is an odd number.

Look at each of these expressions and decide whether it is *always even*, *always odd* or can be *either odd or even*. Tick the correct box.

	Always even	Always odd	Either odd or even	
a $P + S$	☐	☐	☐	**[1 mark]**
b $P \times S$	☐	☐	☐	**[1 mark]**
c Q^2	☐	☐	☐	**[1 mark]**
d $P + Q$	☐	☐	☐	**[1 mark]**

AU 3   Here are three numbers.

     20   25   36

Give a reason why each of them could be the odd one out.

                          _____

                          _____

                          _____

                                 [3 marks]

1   Here are six number cards.

a   Use **three** of the cards to make this statement true.

$$\sqrt{\boxed{}\boxed{}} = \boxed{}$$

[1 mark]

b   Use **three** different cards to make this statement true.

$$\sqrt{\boxed{}\boxed{}} = \boxed{}$$

[1 mark]

c   Which is greater, $\sqrt{144}$ or 2^4? _____ **[1 mark]**

d   Write down the value of each number.

   i $\sqrt{169}$ **ii** 5^3

_____   _____ **[1 mark each]**

e   Write down the value of each number.

   i $\sqrt[3]{64}$ **ii** 2^8

_____   _____ **[1 mark each]**

2 a   Fill in the missing numbers.

			Units digit
4^1	=	4	4
4^2	=	16	6
4^3	=	——	——
4^4	=	——	——
4^5	=	——	——

[2 marks]

b   What is the last digit of 4^{99}? Explain your answer.

_____ **[1 mark]**

c   Which is greater, 5^6 or 6^5? Justify your answer.

_____ **[1 mark]**

AU 3   Arrange these numbers in order, starting with the smallest

$\sqrt[3]{216}$ 2^3 $\sqrt{25}$

_____ **[2 marks]**

Powers of 10

1 a What is the value of the digit **4** in the number 23.4? _____ **[1 mark]**

b Write 10 000 in the form 10^n, where n is an integer. _____ **[1 mark]**

c Write, in full, the number represented by 10^7. _____ **[1 mark]**

d Fill in the missing numbers and powers.

1000	☐	10	1	$\frac{1}{10}$	☐	$\frac{1}{1000}$
10^3	10^2	$10^{☐}$	$10^{☐}$	10^{-1}	10^{-2}	$10^{☐}$

[1 mark]

e Write down the value of:

i 6^0 _____ **[1 mark]**

2 a Work out:

i 3.7×10^2 _____ **[1 mark]**

ii 0.25×10^3 _____ **[1 mark]**

b Work out:

i $7.6 \div 10$ _____ **[1 mark]**

ii $0.65 \div 10^2$ _____ **[1 mark]**

c Work out:

i $30\,000 \times 400$ _____ **[1 mark]**

ii 600^2 _____ **[1 mark]**

d Work out:

i $90\,000 \div 30$ _____ **[1 mark]**

ii $30\,000 \div 60$ _____ **[1 mark]**

AU 3 Use the following diagrams to work out how many cubic centimetres (cm³) there are in a cubic metre (m³).

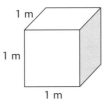

_____ **[2 marks]**

C

1 a What number is represented by $2 \times 3^2 \times 5$? _____ **[1 mark]**

   b Write 70 as the product of its prime factors. _____ **[1 mark]**

   c Write 48 as the product of its prime factors. _____ **[1 mark]**

   d i Complete the prime factor tree to find the prime factors of 900.

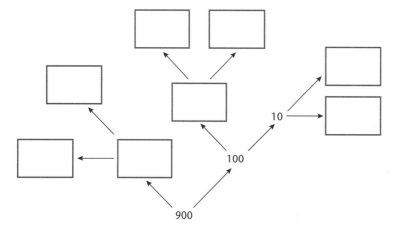

[2 marks]

   ii Write 900 as the product of its prime factors, in index form.

_____ **[1 mark]**

C

2 a i You are given that $3x^2 = 75$.

      What is the value of x?

      $x =$ _____ **[1 mark]**

   ii Write 150 as the product of its prime factors.

      $x =$ _____ **[1 mark]**

   b i You are given that $2x^3 = 54$.

      What is the value of x?

      $x =$ _____ **[1 mark]**

   ii Write 216 as the product of its prime factors.

      $x =$ _____ **[1 mark]**

C

PS 3 a and b are prime numbers, such that $a^2b = 2009$.

Find the values of a and b.

_____ **[3 marks]**

LCM and HCF

AQA 2   EDEXCEL 2   OCR 1

1 a Write 24 as the product of its prime factors. _____ **[1 mark]**

b Write 60 as the product of its prime factors. _____ **[1 mark]**

c What is the lowest common multiple of 24 and 60? _____ **[1 mark]**

d What is the highest common factor of 24 and 60? _____ **[1 mark]**

e In prime factor form, the number $P = 2^4 \times 3^2 \times 5$ and the number $Q = 2^2 \times 3 \times 5^2$.

    i What is the lowest common multiple of P and Q?

    Give your answer in index form. _____ **[1 mark]**

    ii What is the highest common factor of P and Q?

    Give your answer in index form. _____ **[1 mark]**

---

2 a You are told that p and q are prime numbers.

$p^2q^2 = 36$

What are the values of p and q?

$p = $ _____

$q = $ _____ **[2 marks]**

b Write 360 as the product of its prime factors.

_____ **[1 mark]**

(PS c) You are told that a and b are prime numbers.

$ab^2 = 98$

What are the values of a and b?

$a = $ _____

$b = $ _____ **[2 marks]**

d Write 196 as the product of its prime factors.

_____ **[1 mark]**

(AU 3) a i Find the HCF of 3 and 17.

_____ **[1 mark]**

    ii Find the LCM of 3 and 17.

_____ **[1 mark]**

b p and q are prime numbers.

    i Write down the HCF of p and q.

_____ **[1 mark]**

    ii Write down the LCM of p and q.

_____ **[1 mark]**

C

1 a Write $4^3 \times 4^5$ as a single power of 4.

_____ **[1 mark]**

b Write $6^5 \div 6^2$ as a single power of 6.

_____ **[1 mark]**

c i If $3^n = 81$, what is the value of n?

_____ **[1 mark]**

ii If $3^m = 27$, what is the value of m?

_____ **[1 mark]**

d Write down the actual value of $7^9 \div 7^7$.

_____ **[1 mark]**

e Write down the actual value of $10^4 \times 10^4$.

_____ **[1 mark]**

f Write down the actual value of $2^2 \times 5^2 \times 2^4 \times 5^4$.

_____ **[1 mark]**

C

2 a Write $x^5 \times x^2$ as a single power of x.

_____ **[1 mark]**

b Write $x^8 \div x^4$ as a single power of x.

_____ **[1 mark]**

c $2^3 \times 3^3 = 6^3$.

Which of the following expressions is the same as $a^n \times b^n$?

$(a + b)^n$ ab^{2n} $(ab)^n$

_____ **[1 mark]**

d $8^3 \div 2^3 = 4^3$.

Which of the following expressions is the same as $a^n \div b^n$?

$(a \div b)^n$ $a \div b^n$ $a^n - b^n$

_____ **[1 mark]**

C

PS 3 a Use trial and improvement to find the value of a such that $(1 + a)^2 = a^2$.

_____ **[2 marks]**

b Is the following rule true?

$(1 + a)^3 = 1 + 3a + 3a^2 + a^3$

☐ Yes   ☐ No

Justify your choice.

_____ **[2 marks]**

1 a Arne uses the column method to work out 37×48. This is his working.

```
        3    7
    ×   4    8
    ─────────
    2   9₅   6
    1   4₂   8
    ─────────
    4   4    4
    1    1
```

Arne has made a mistake.

   i What mistake has Arne made?

_____   **[1 mark]**

   ii Work out the correct answer to 37×48.   _____   **[1 mark]**

b Berne is using the box method to work out 29×47. This is his working.

×	20	9
40	60	49
7	27	16

```
        6    0
        4    9
        2    7
    +   1    6
    ─────────
    1   5    2
```

Berne has made a mistake.

   i What mistake has Berne made?

_____   **[1 mark]**

   ii Work out the correct answer to 29×47.

_____   **[2 marks]**

2 a There are 144 plasters in a box. How many plasters will there be in 24 boxes?

_____   **[2 marks]**

  b Each box of plasters costs 98p. How much will 24 boxes of plasters cost?

    Give your answer in pounds and pence.   _____   **[2 marks]**

3 Here is part of the 29 times table.

  ×1   ×2   ×3 ×4   ×5

  29   58   87   116   145

Use the table above to work these out.

a 9×29 _____

b 52×29 _____

c 105×29 _____

[3 marks]

1 a Write down the answers to these.

    i $1 \times 29 =$ _____

    ii $2 \times 29 =$ _____

    iii $10 \times 29 =$ _____

    iv $20 \times 29 =$ _____ **[2 marks]**

b Using the values above, or otherwise, complete
this division by the chunking method.

    $1508 \div 29$ 1   5   0   8

                      –   5   8   0

                            ——————

[2 marks]

c **i** Write down the answer to $1520 \div 29$.   _____ **[1 mark]**

    ii Write down the answer to $1508 \div 58$.   _____ **[1 mark]**

2 a A widget machine produces 912 widgets per hour.

How many widgets does it produce during a 16-hour shift?

_____ **[2 marks]**

b The widgets are packed in boxes of 24.

How many boxes will be needed to pack 912 widgets?

_____ **[2 marks]**

(PS 3) You are given that $1248 \div 52 = 24$.

Write down the answer to each of these.

    a $624 \div 52$ _____ **[1 mark]**

    b $1248 \div 26$ _____ **[1 mark]**

    c $1248 \div 12$ _____ **[1 mark]**

    d $2496 \div 6$ _____ **[1 mark]**

1 a The school canteen has 34 tables.

Each table can seat up to 14 students.

What is the maximum number of people who could use the canteen at one time?

_____ **[2 marks]**

b The head wants to expand the canteen so it can cater for 600 students.

How many extra tables will be needed?

_____ **[2 marks]**

2 a This table shows the column headings for the number 23.4789.

10	1	•	$\frac{1}{10}$	$\frac{1}{100}$	$\frac{1}{1000}$	$\frac{1}{10\,000}$
2	3	•	4	7	8	9

i The number 23.4789 is multiplied by 100.

What will be the place value of the digit 4 in the **answer** to 23.4789 × 100? _____ **[1 mark]**

ii The number 23.4789 is divided by 10.

What will be the place value of the digit 7 in the **answer** to 23.4789 ÷ 10? _____ **[1 mark]**

b Round the number 23.4789 to:

i   1 decimal place _____ **[1 mark]**

ii 2 decimal places _____ **[1 mark]**

iii 3 decimal places. _____ **[1 mark]**

(PS 3) Kieron knows that four decimal numbers have units of 1, 2, 3 and 7 and decimal parts of 0.08, 0.24, 0.86 and 0.93 but he cannot remember which unit goes with which decimal.

He knows that the six possible totals when any two are added together are

11.10, 9.94, 9.79, 5.32, 5.17 and 4.01

Match up the units with the appropriate decimal.

_____ **[3 marks]**

E

1 a Complete the shopping bill.

3 jars of jam at £1.28 per jar	
2 kg of apples at £2.15 per kg	
5 doughnuts at 22p each	
Total	**[4 marks]**

b Work these out.

i 3×2.6

_____ **[1 mark]**

ii 2.4×2.6

_____ **[2 marks]**

c A dividend is a repayment made every few months on the amount spent.
The Co-op pays a dividend of 2.6 pence for every £1 spent.

i In three months, Derek spends £240.

How much dividend will he get? _____ **[1 mark]**

ii Doreen received a dividend of £7.80.

How much did she spend to get this dividend? _____ **[1 mark]**

F

2 a i Work out 6×2.9.

_____ **[1 mark]**

ii Work out 4.6×2.9.

_____ **[2 marks]**

b What is the cost of 2.9 kg of coffee beans at £4.60 per kilogram?

_____ **[1 mark]**

C

(PS 3) Bernard went to the USA for three months between January and April.

When he went he changed £10 000 into dollars at an exchange rate of £1 = $1.35.

He spent $11 000 and changed the remaining dollars back in April at an exchange rate of £1 = $1.65.

How many pounds did he get back?

_____ **[3 marks]**

More fractions

1   **a**   Work out $\frac{3}{4} + \frac{2}{5}$.

Give your answer as a mixed number.

_____ **[2 marks]**

b   Work out $3\frac{2}{3} - 1\frac{4}{5}$.

Give your answer as a mixed number.

_____ **[3 marks]**

c   On an aeroplane, two-fifths of the passengers were British, one-quarter were German, one-sixth were American and the rest were French.

What fraction of the passengers are French?

_____ **[3 marks]**

2   **a**   Work out $2\frac{1}{2} \times 1\frac{2}{5}$.

_____ **[2 marks]**

b   Work out $3\frac{3}{10} \div 2\frac{2}{5}$.

Give your answer as a mixed number.

_____ **[3 marks]**

c   Work out the area of this triangle.

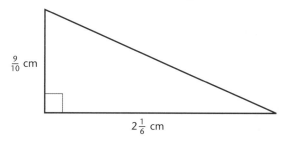

$\frac{9}{10}$ cm

$2\frac{1}{6}$ cm

cm^2

_____ **[2 marks]**

PS 3   My son is $\frac{3}{5}$ of my age.

My daughter is $\frac{2}{3}$ of my age

Altogether the sum of our ages is 136 years.

How old am I?

_____ **[3 marks]**

1 a Complete the following calculations.

i +6 × −2 = ☐ **[1 mark]**

ii −4 × ☐ = 12 **[1 mark]**

iii ☐ ÷ −5 = +4 **[1 mark]**

b Complete this sentence, using two negative numbers.

 − = +2 **[1 mark]**

c Look at these numbers.

−4, −3, −2, 0, 2, 5, 9

i Mazy is asked to pick out the square numbers. She chooses −4 and 9.

Explain why Mazy is wrong.

_____ **[1 mark]**

ii Work out the mean of the numbers.

_____
_____ **[2 marks]**

---

2 a Round each of these numbers to one significant figure.

i 52.1   ii 0.38

_____   _____ **[1 mark each]**

b Find an approximate value for $\dfrac{52.1 \times 39.6}{18.7 - 11.1}$.

_____ **[1 mark]**

c Find an approximate value for $\dfrac{52.1 - 39.6}{0.38}$.

_____ **[2 marks]**

AU 3 **a Write three numbers, two positive and one negative, in the boxes to make the equation true.**

☐ × ☐ ÷ ☐ = −2 **[1 mark]**

b Write three negative numbers in the boxes to make the equation true.

☐ × ☐ ÷ ☐ = −2 **[1 mark]**

C

1 The angles of a quadrilateral are in the ratio 2:3:5:8.

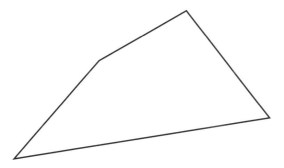

What is the value of the largest angle?

_____ °   **[3 marks]**

C

2 a Write the ratio 12:9 in its simplest form.

_____   **[1 mark]**

b Write the ratio 5:2 in the form 1:n.

_____   **[1 mark]**

c A fruit drink is made from orange juice and cranberry juice in the ratio 5:3.

If 1 litre of the drink is made, how much of the drink is cranberry juice?

_____   **[2 marks]**

C

3 In a tutor group the ratio of boys to girls is 3:4.

There are 15 boys in the form.

How many students are there in the form altogether?

_____   **[2 marks]**

C

AU 4 The ratio of pine trees to oak trees in a wood is 3 : 5.

Are the following statements True (T), False (F) or could be true (C)?

Put ticks in the appropriate boxes.

The first is done for you.

Statement	T	F	C
There are 25 pine trees in the wood.		✓	
There are 800 trees altogether in the wood.			
The fraction of pine trees in the wood is $\frac{3}{5}$.			
The percentage of oak trees in the wood is 60%.			
If half of the pine trees were cut down the ratio of pine trees to oak trees would now be 3 : 10.			

[3 marks]

D

1 A ferry covers the 72 kilometres between Holyhead and Dublin in $2\frac{1}{4}$ hours.

   a What is the average speed of the ferry?

      State the units of your answer.

_____ **[3 marks]**

   b For the first 15 minutes and the last 15 minutes, the Ferry is manoeuvring in and out of the ports. During this time the average speed is 18 km per hour.

      What is the average speed of the ferry during the rest of the journey?

_____ **[3 marks]**

D

2 A car uses 50 litres of petrol in driving 275 miles.

   a How much petrol will the car use in driving 165 miles?

_____ **[2 marks]**

   b How many miles can the car drive on 26 litres of petrol?

_____ **[2 marks]**

D

3 Nutty Flake cereal is sold in two sizes.

The handy size contains 600 g and costs £1.55.

The large size contains 800 g and costs £2.20.

Which size is the better value?

_____ **[3 marks]**

C

(**PS 4**) Josh cycles 20 km up a very steep hill at 10 km/h.

He turns round at the top and cycles back at 40 km/h.

His wife says that his average speed must be $\frac{10 + 40}{2} = 25$ km/h.

Is Josh's wife correct?

Justify your answer.

_____ **[3 marks]**

1 **a** Complete this table of equivalent fractions, decimals and percentages.

Decimal	Fraction	Percentage
0.35		
	$\frac{4}{5}$	
		90%

[3 marks]

b **i** What fraction of this diagram is shaded?

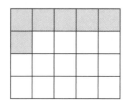

Give your answer in its simplest form. _____ [1 mark]

ii What percentage is **not** shaded?

_____ [1 mark]

c Write these values in order of size, smallest first.

0.6 $\frac{11}{20}$ 57% $\frac{14}{25}$

_____ [2 marks]

2 **a** A car cost £6000 new.

It depreciated in value by 12% in the first year.

It then depreciated in value by 10% in the second year.

Which of these calculations shows the value of the car after two years?

Circle the correct answer.

6000×0.78 $6000 \times 0.88 \times 0.9$ $6000 \times 88 \times 90$ $6000 - 2200$ [1 mark]

b VAT is charged at $17\frac{1}{2}$%.

A quick way to work out the VAT on an item is to work out 10%, then divide this by 2 to get 5%, then divide this by 2 to get $2\frac{1}{2}$%. Add these values all together.

Use this method to work out the VAT on an item costing £68.

_____ [2 marks]

AU 3 Shop A increased its prices by 5%, then reduced them by 3%.

Shop B reduced its prices by 3%, then increased them by 5%.

Which shop's prices were the greatest after the changes?

Circle the correct answer

Shop A   Shop B   Both same   Cannot tell

Justify your choice.

_____ [3 marks]

E

1 A washing machine normally costs £350.

Its price is reduced by 15% in a sale.

   a What is 15% of £350?

£ _____ **[2 marks]**

   b What is the sale price of the washing machine?

£ _____ **[1 mark]**

D

2 **a** A computer costs £700, not including VAT.

VAT is charged at $17\frac{1}{2}\%$.

What is the cost of the computer when VAT is added?

£ _____ **[2 marks]**

   b The price of a printer is reduced by 12% in a sale.

The original price of the printer was £250.

What is the price of the printer in the sale?

£ _____ **[2 marks]**

C

3 In the first week it was operational, a new bus route carried a total of 2250 people.

In the second week it carried 2655 people.

What is the percentage increase in the number of passengers from the first week to the second?

_____ **[3 marks]**

C

AU 4 Explain why a 10% increase followed by a 10% decrease does not get back to the original amount.

_____

_____ **[2 marks]**

Number record sheet

Name _____   **Marks** _____

Form _____   **Percentage** _____

Date _____   **Grade** _____

Page number	Question number	Topic	Mark	Comments
124	1	Basic number	/5	
	2	Basic number	/3	
	3	Basic number	/4	
125	1	Basic number	/3	
	2	Basic number	/7	
	3	Basic number	/3	
	4	Basic number	/2	
	5	Basic number	/3	
126	1	Basic number	/4	
	2	Basic number	/6	
	3	Basic number	/5	
	4	Basic number	/3	
127	1	Fractions	/4	
	2	Fractions	/3	
	3	Fractions	/2	
	4	Fractions	/3	
128	1	Fractions	/11	
	2	Fractions	/8	
	3	Fractions	/3	
129	1	Fractions	/12	
	2	Fractions	/7	
	3	Fractions	/3	
130	1	Fractions	/8	
	2	Fractions	/4	
	3	Fractions	/4	
	4	Fractions	/3	
131	1	Rational numbers	/8	
	2	Rational numbers	/4	
	3	Rational numbers	/3	
132	1	Negative numbers	/4	
	2	Negative numbers	/5	
	3	Negative numbers	/2	
133	1	Negative numbers	/6	
	2	Negative numbers	/2	
	3	Negative numbers	/2	
134	1	More about number	/4	
	2	More about number	/5	
	3	More about number	/1	
135	1	Primes and squares	/6	
	2	Primes and squares	/4	
	3	Primes and squares	/3	
136	1	Roots and powers	/7	
	2	Roots and powers	/4	
	3	Roots and powers	/2	

Page number	Question number	Topic	Mark	Comments
139	1	Powers of 10	/6	
	2	Powers of 10	/8	
	3	Powers of 10	/2	
140	1	Prime factors	/6	
	2	Prime factors	/4	
	3	Prime factors	/3	
141	1	LCM and HCF	/6	
	2	LCM and HCF	/6	
	3	LCM and HCF	/4	
142	1	Powers	/7	
	2	Powers	/4	
	3	Powers	/4	
143	1	Number skills	/5	
	2	Number skills	/4	
	3	Number skills	/3	
144	1	Number skills	/6	
	2	Number skills	/4	
	3	Number skills	/4	
145	1	Number skills	/4	
	2	Number skills	/5	
	3	Number skills	/3	
146	1	Decimals	/9	
	2	Decimals	/4	
	3	Decimals	/3	
147	1	More fractions	/8	
	2	More fractions	/7	
	3	More fractions	/3	
148	1	More number	/7	
	2	More number	/5	
	3	More number	/2	
149	1	Ratio	/3	
	2	Ratio	/4	
	3	Ratio	/2	
	4	Ratio	/3	
150	1	Speed and proportion	/6	
	2	Speed and proportion	/4	
	3	Speed and proportion	/3	
	4	Speed and proportion	/3	
151	1	Percentages	/7	
	2	Percentages	/3	
	3	Percentages	/3	
152	1	Percentages	/3	
	2	Percentages	/4	
	3	Percentages	/3	
	4	Percentages	/2	
Total			**/394**	

Successes

1 _____

2 _____

3 _____

Areas for improvement

1 _____

2 _____

3 _____

1   Here is a rectangle.

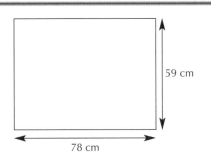

59 cm

78 cm

a   Find the perimeter of the rectangle.

_____

[2 marks]

b   Find the area of the rectangle.

_____ **[2 marks]**

2   **a**   The map of an island is drawn on the centimeter-square grid.

The scale is 1 cm represents 1 km.

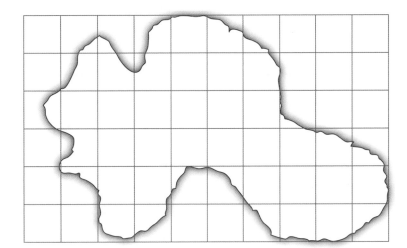

Estimate the area of the island.

_____ **[2 marks]**

b   The diagram shows a trapezium drawn on a centimetre-square grid.

By counting squares, or otherwise, find the area of the trapezium.

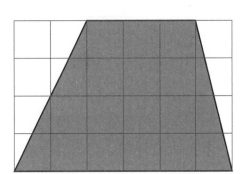

cm² **[2 marks]**

PS 3   This square has the same numerical value for its perimeter and its area.

x

What is the value of x?

_____ **[2 marks]**

D

1 Calculate the area of this shape.

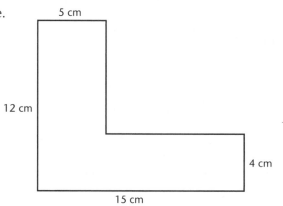

_____ cm² **[3 marks]**

D

2 a Calculate the area of this triangle.

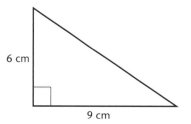

_____ cm² **[2 marks]**

b Calculate the area of this shape.

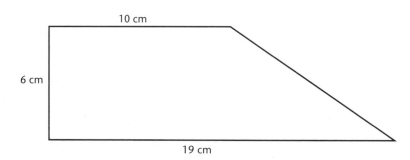

_____ cm² **[2 marks]**

D

PS 3 The area of this L shape is 39 cm².

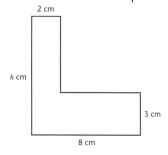

Work out the length h. _____ **[3 marks]**

1 a The parallelogram is drawn on a centimetre-square grid.

Find the area of the parallelogram.

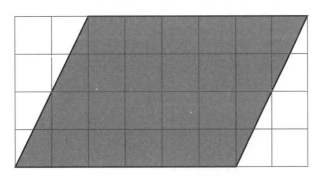

_____ cm² **[2 marks]**

b Calculate the area of this parallelogram.

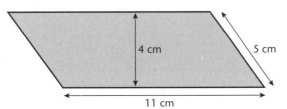

4 cm

5 cm

11 cm

_____ cm² **[2 marks]**

2 Calculate the area of this trapezium.

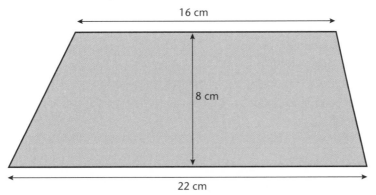

16 cm

8 cm

22 cm

_____ cm² **[2 marks]**

AU 3 The area of a trapezium is given on the formula sheet.

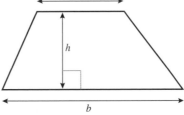

Area of trapezium = $\frac{1}{2}(a + b)h$

a

h

b

Find values of a, b and h so that the area of the trapezium is 60 cm².

_____ **[2 marks]**

G 1

a Which of the letters above have line symmetry? _____ **[2 marks]**

b Which of the letters above have rotational symmetry of order 2?

_____ **[1 mark]**

c

    i How many lines of symmetry does the parallelogram have?

_____ **[1 mark]**

    ii What is the order of rotational symmetry of a parallelogram?

_____ **[1 mark]**

F 2 a Shade in **five** more squares so that the pattern has no lines of symmetry and rotational symmetry of order 2.

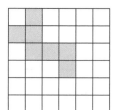

[1 mark]

b The diagram shows an octahedron made of two equal square-based pyramids joined by their bases.

How many planes of symmetry does this octahedron have?

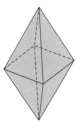

_____ **[1 mark]**

E (PS 3) The diagram shows a regular octagon split into 8 congruent triangles.

This has 8 lines of symmetry and rotational symmetry order 8.

This can be denoted by 8L, 8R.   This diagram is 1L, 0R.

a Shade three triangles in this octagon so it is 0L, 0R. **[1 mark]**

b Shade four triangles in this octagon so that it is 2L, 2R. **[1 mark]**

1 a Measure the following angles.

i

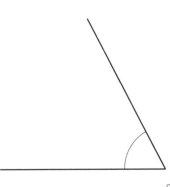

ii

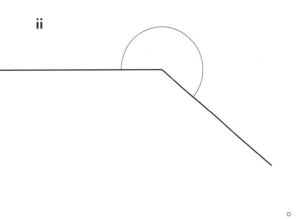

_____ °   _____ ° **[2 marks]**

b Draw an angle of 55°.

[1 mark]

2 a The diagram shows three angles on a straight line.

Work out the value of x.

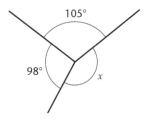

_____ ° **[2 marks]**

b The diagram shows three angles meeting at a point.

Work out the value of x.

105°

98°

x

_____ ° **[2 marks]**

PS 3 Ashley draws three angles meeting at a point.

One angle is acute, one angle is obtuse and one angle is reflex.

Write down three values that the angles could be.

Acute _____ Obtuse _____ Reflex _____ **[3 marks]**

F

1 a This triangle has two equal sides.

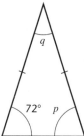

i   What name is given to this type of triangle? _____ **[1 mark]**

ii Find the values of p and q.

$p =$ _____ ° **[1 mark]**

$q =$ _____ ° **[1 mark]**

b Work out the size of the angle marked x.

_____ ° **[2 marks]**

E

2 The diagram shows a kite.

Work out the size of the angle marked y.

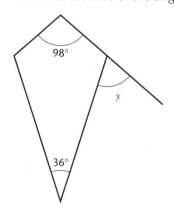

_____ ° **[3 marks]**

C

PS 3 All the angles in this kite are multiples of 10.

The angle at A is 100°.

What is the largest value that D and B could be? _____ **[3 marks]**

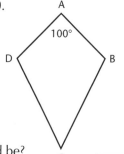

Polygons

D

1 a The diagram shows a regular octagon.

O is the centre of the octagon.

Calculate the sizes of angles p, q and r.

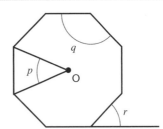

$p =$ _____ °

$q =$ _____ °

$r =$ _____ °   **[3 marks]**

b Explain why the interior angles of a pentagon add up to 540°.

_____

_____

_____   **[2 marks]**

2 a The diagram shows three sides of a regular polygon.

The exterior angle is 36°.

How many sides does the polygon have altogether?

_____   **[2 marks]**

C

b The interior angle of a regular polygon is 160°.

Explain why the polygon must have 18 sides.

_____

_____

_____   **[2 marks]**

PS 3 A hexagon and a pentagon with equal sides are placed together.

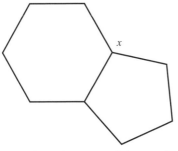

Calculate the size of the angle x between the sides.   _____   **[3 marks]**

C

C

1 In the diagram, QR is parallel to LM.

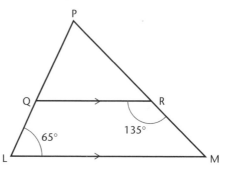

a Write down the size of angle RML.

Give a reason for your answer.

RML = _____ ° **[1 mark]**

Reason: _____ **[1 mark]**

b Write down the size of angle PQR.

Give a reason for your answer.

PQR = _____ ° **[1 mark]**

Reason: _____ **[1 mark]**

c Work out the size of angle QPR.

QPR = _____ ° **[1 mark]**

C

2 The lines AB and CD are parallel.

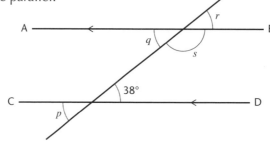

Write down the sizes of angles p, q, r and s, and in each case give the reason in relation to the given angle of 38°.

p = _____° because it is _____ to the given angle of 38°. **[2 marks]**

q = _____° because it is _____ to the given angle of 38°. **[2 marks]**

r = _____° because it is _____ to the given angle of 38°. **[2 marks]**

s = _____° because it is _____ to the given angle of 38°. **[2 marks]**

C **AU 3** In the diagram, is the line AB parallel to the line CD?

Give a reason for your answer,
referring to angles on the diagram.

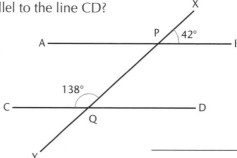

_____ **[2 marks]**

1 a Jonathan is describing a quadrilateral.

What quadrilateral is he describing?

It has no lines of symmetry and has rotational symmetry of order 2.

The diagonals cross at right angles and all the sides are equal.

_____ **[1 mark]**

b Marie is describing another quadrilateral.

  i Write down the name of a quadrilateral that Marie could be describing.

_____ **[1 mark]**

  ii Write down the name of a different quadrilateral that Marie could be describing.

_____ **[1 mark]**

2 The diagram shows a kite, ABCD, attached to a parallelogram, CDEF.

Angle BAD = 100°, angle CFE = 60°.

When the side of the kite is extended it passes along the diagonal of the parallelogram.

Use the properties of quadrilaterals to work out the size of angle CED.

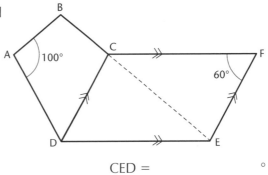

CED = _____ ° **[4 marks]**

AU 3 These are the standard quadrilaterals

Square   Rectangle   Rhombus   Kite   Parallelogram   Trapezium

All quadrilaterals have four sides.

a State a mathematical property that the **square** and the **rectangle** have in common.

_____ **[1 mark]**

b State a mathematical property that the **kite** and the **rhombus** have in common.

_____ **[1 mark]**

c State a mathematical property that the **parallelogram** and the **rectangle** have in common.

_____ **[1 mark]**

d Choose any other pair of quadrilaterals and write down a mathematical property that they have in common

Quadrilateral 1   _____

Quadrilateral 2   _____

Property   _____

[1 mark]

D

1 This map shows the positions of four towns, Althorp, Beeton, Cowton and Deepdale. The scale is 1 cm to 1 kilometre.

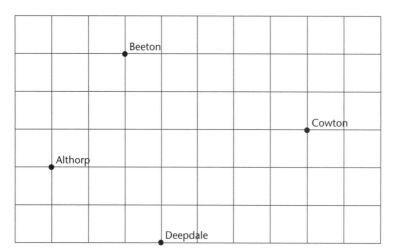

Using a ruler and protractor, find the distance and bearing of:

a Beeton from Althorp   _____ km at _____ ° **[2 marks]**

b Deepdale from Beeton   _____ km at _____ ° **[2 marks]**

c Deepdale from Cowton   _____ km at _____ ° **[2 marks]**

d Althorp from Deepdale   _____ km at _____ ° **[2 marks]**

C

2 Brian walks in a perfect square.

He starts by walking north for 100 m and then turning right.

He then continues walking for 100 m then turning right, doing it twice more until he is back to his starting point.

Write down the four bearings of the directions in which he walks.

_____ **[4 marks]**

C

PS 3   **a** Measure the bearings of A to B and B to A.   _____ **[1 mark]**

b Measure the bearings of C to D and D to C.   _____ **[1 mark]**

c What is the connection between the two bearings in **a** and **b**?   _____ **[1 mark]**

1 From this list of words, fill in the missing words that describe parts of the circle on the diagram.

Chord   Tangent   Radius   Diameter

O is the centre of the circle.

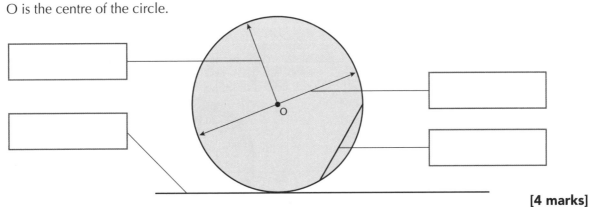

[4 marks]

2 a Work out the area of a circle of radius 12 cm.

Give your answer to 1 decimal place.

_____ cm² **[2 marks]**

b Work out the circumference of a circle of diameter 20 cm.

Give your answer to 1 decimal place.

_____ cm **[2 marks]**

3 Work out the area of a semicircle of diameter 20 cm.

Give your answer in terms of π.

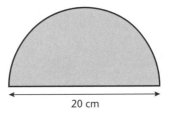

20 cm

_____ cm² **[2 marks]**

PS 4 A circle with a radius of 4 cm has another circle of radius *r* cut from it.

The area of the ring that is left is equal to the area of the hole in the centre.

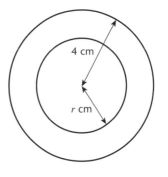

4 cm

r cm

What is the radius of the inner circle?   _____ **[3 marks]**

G

1 a The thermometer shows the temperature outside Frank's local garage on a hot summer day. What temperature does the thermometer show?

_____ °C **[1 mark]**

b Later, when Frank was driving home, the speedometer looked like this.

What speed was Frank doing?

State the units of your answer.

_____ **[2 marks]**

c At the same time the car rev counter showed 3300 rpm.

Draw an arrow to show 3300 rpm on the scale.

[1 mark]

F

AU 2 The picture shows a coastguard with a beached whale.

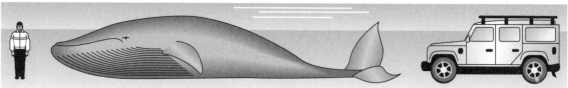

Estimate the length of the whale.

Give your answer in metres.

_____ m **[2 marks]**

D

PS 3 Here are two scales.

Mark the point on the scales where the reading on both will be the same and in line vertically.

[2 marks]

1   The diagram shows a line AB and a point C, drawn on a centimetre-square grid.

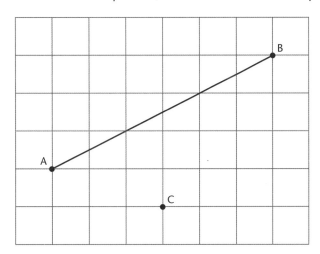

a   Measure the length of the line AB, in centimetres. _____ cm **[1 mark]**

b   Mark the midpoint of AB with a cross. **[1 mark]**

c   Draw a line through the point C, perpendicular to the line AB. **[1 mark]**

d   The diagram represents a map with a scale of 1 cm to 5 km.

Work out the real distance represented by BC.

_____ km **[2 marks]**

2   The net of a solid is shown, drawn to scale.

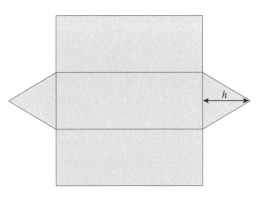

a   What is the name of the solid for which this is the net?

_____ **[1 mark]**

b   Measure the height of the triangle, h, shown on the net.

_____ cm **[1 mark]**

c   By taking appropriate measurements, work out the surface area of the net.

_____ cm^2 **[3 marks]**

PS 3   On a map drawn to a scale of 1 cm to 20 m a building is 2 cm by 3 cm.

The area of the building on the map is 6 cm^2.

What is the area of the actual building?

_____ **[2 marks]**

1 The diagram shows a prism, with a T-shaped cross-section, drawn on a one-centimetre isometric grid.

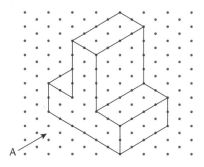

A

a What is the volume of the prism? _____ cm³   **[1 mark]**

b Draw the side elevation of the prism, from A.

[1 mark]

PS 2 The diagram shows the plan and both side elevations of a solid made from five one-centimetre cubes.

Draw an isometric view of the solid on the grid below.

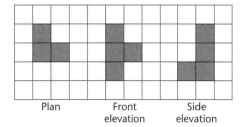

Plan   Front   Side
          elevation elevation

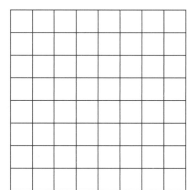

[2 marks]

PS 3 This is the net of a solid.

Draw an isometric view of the solid on the grid below.

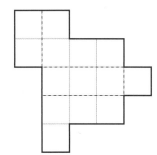

[2 marks]

Congruency and tessellations

1 The grid shows six shapes A, B, C, D, E and F.

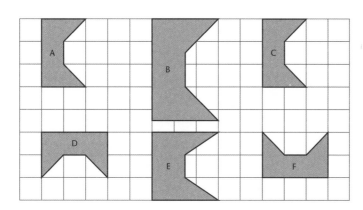

a Write down the letters of the shapes that are congruent to shape A.

_____ **[1 mark]**

b Which shape is similar to shape A?

_____ **[1 mark]**

2 a The diagram shows a tessellation of an isosceles triangle.

Explain what is meant by a *tessellation*.

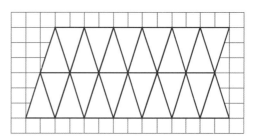

_____

_____ **[1 mark]**

b Use the shape shown to draw a tessellation.

Draw at least six more shapes to show the tessellation clearly.

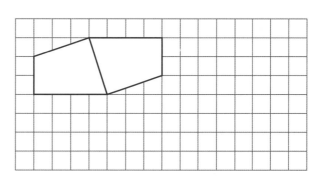

[1 mark]

(**AU 3**) Explain, using a diagram, why a regular octagon will never tessellate.

[2 marks]

Geometry

C

1 a Describe the transformation that takes the shaded triangle to triangle A.

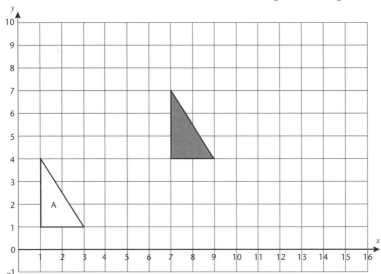

_____

_____ **[2 marks]**

b Translate the shaded triangle by $\binom{-4}{2}$. Label the image B. **[1 mark]**

c The shaded triangle is translated by $\binom{7}{-5}$ to give triangle C.

What **vector** will translate triangle C to the shaded triangle? _____ **[1 mark]**

D

2

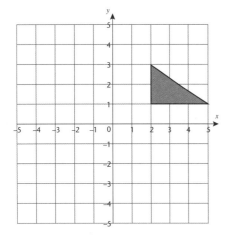

a Reflect the shaded triangle in the y-axis. Label the image A. **[1 mark]**

b Reflect the shaded triangle in the line $y = -1$. Label the image B. **[1 mark]**

c Reflect the shaded triangle in the line $y = -x$. Label the image C. **[1 mark]**

C

PS 3 Square A can be transformed to square B by a translation and a reflection.

Describe each of these.

Translation of $\begin{pmatrix} \text{.......} \\ \text{.......} \end{pmatrix}$ **[1 mark]**

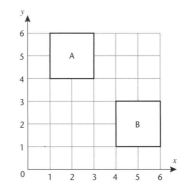

Reflection in _____ **[1 mark]**

1

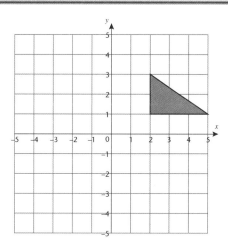

D

a   Rotate the shaded triangle by 90° clockwise about (1, 0). Label the image A.   **[1 mark]**

b   Rotate the shaded triangle by a half-turn about (1, 3). Label the image B.   **[1 mark]**

c   What rotation will take triangle A to triangle B?

_____   **[3 marks]**

2

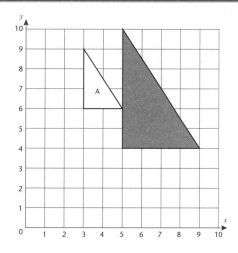

D

a   What transformation takes the shaded triangle to triangle A?

_____   **[2 marks]**

b   Draw the image after the shaded triangle is enlarged by a scale
factor $\frac{1}{4}$ centred on (1, 0).   **[2 marks]**

PS 3   Triangle A is an enlargement of the shaded triangle,
scale factor $1\frac{1}{2}$, centre (1, 0).

Triangle B is an enlargement of triangle A, scale
factor 2, centre (1, 6).

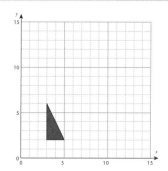

C

What transformation takes the shaded triangle to
triangle B?   _____   **[3 marks]**

1 a Make an accurate drawing of this triangle.

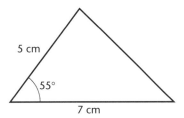

[3 marks]

b Use compasses and a ruler to construct this triangle accurately.   **[3 marks]**

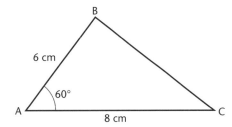

2 Use compasses and a ruler to do these constructions.

a Construct the perpendicular bisector of AB.

A ●

● B

[2 marks]

Use compasses and ruler to do these constructions.

b Construct the perpendicular bisector of the line L.

——————————————— L   **[2 marks]**

AU 3 Bashir and Mona are constructing triangles.

Bashir

> My triangle has one side 7 cm long, one side 8 cm long and one angle of 40°.

Mona

> My triangle also has one side 7 cm long, one side 8 cm long and one angle of 40°, so it must be congruent to yours.

Sketch two triangles, marking on the lengths and angles to show that Mona is wrong.

[3 marks]

1 Construct the bisector of angle ABC.

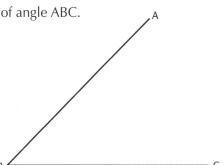

[2 marks]

2 The diagram, which is drawn to scale, shows a flat, rectangular lawn of length 10 m and width 6 m, with a circular flower bed of radius 2 m.

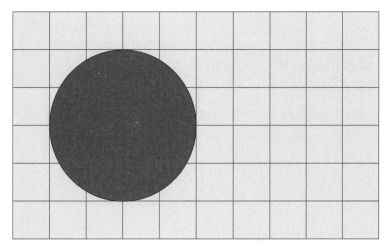

Scale: 1 cm represents 1 m.

A tree is going to be planted in the garden.

It has to be at least 1 metre from the edge of the garden and at least 2 metres from the flower bed.

a Draw a circle to show the area around the flower bed where the tree *cannot* be planted. **[1 mark]**

b Show the area of the garden in which the tree *can* be planted. **[2 marks]**

PS 3 This is the plan of a paddling pool.

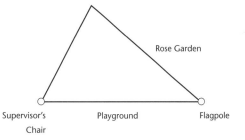

The council want to rope off an area for model boat clubs to use.

They decide this area has to be:

• closer to the flagpole than to the supervisor's chair

• nearer to the rose garden than to the playground.

Using a ruler and a pair of compasses, construct the area where the model boats will be allowed. Mark the region clearly with a R.

[4 marks]

F

1 An old water butt is labelled: 'When full contains 50 gallons'.

Mary has a watering can that holds 9 litres.

a How many centilitres is 9 litres?

_____ cl   **[1 mark]**

b Approximately how many times can Mary fill the watering can from the water butt when it is full?

_____   **[2 marks]**

c A bottle of weedkiller says: 'Mix 200 g with 10 litres of water.'

How many grams of weedkiller will Mary have to mix with 9 litres of water?

_____ g   **[2 marks]**

F

2 a The safety instructions for Ahmed's trailer say:

'Load not to exceed 150 kg.'

Ahmed wants to carry 12 bags of sand, each of which weighs 30 lbs.

Can he carry them safely on the trailer?

_____   **[3 marks]**

b Ahmed has to drive 160 kilometres.

i How many metres is 160 kilometres?

_____ m   **[1 mark]**

ii Approximately how many miles is 160 kilometres?

_____ miles   **[1 mark]**

c Ahmed's car travels 30 miles to the gallon.

His tank contains 20 litres.

Will he have enough fuel in the tank to drive 160 kilometres?

_____   **[3 marks]**

D

(PS 3) Salma is in Spain for Christmas. She wants to cook a turkey she bought in England which is labelled '10 pounds weight'.

The only cookery book she can find gives this recipe.

Cocinar para 15 minutos a 200 °C, entonces cocinar a 160 °C para 30 minutos para cada kilo. Which translates as:

Cook for 15 minutes at 200 °C, then cook at 160 °C for 30 minutes for every kilogram.

How long should Salma cook the turkey for?

_____   **[3 marks]**

Surface area and volume of 3-D shapes

F

1 A rectangle measures 20 centimetres by 30 centimetres.

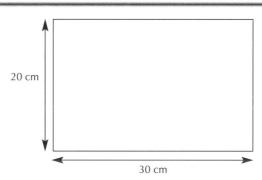

20 cm

30 cm

a Convert 20 centimetres to metres.

_____ m   **[1 mark]**

b Work out the area of the rectangle.

Give your answer in square metres.

_____ m²   **[2 marks]**

D

2 The diagram shows a cuboid, drawn on a centimetre isometric grid.

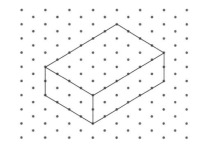

a What are the dimensions of the cuboid?

_____   **[1 mark]**

b Work out the surface area of the cuboid.

_____ cm²   **[2 marks]**

c What is the volume of the cuboid?

_____   **[2 marks]**

C

PS 3 The sides of this cuboid are all whole numbers.

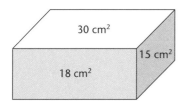

30 cm²

15 cm²

18 cm²

The surface area of each face is shown.

Work out the volume.

_____   **[3 marks]**

Prisms

D

1 A triangular prism has dimensions as shown.

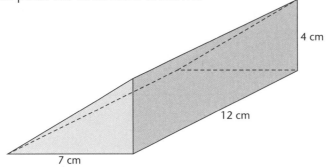

4 cm

12 cm

7 cm

a Calculate the cross-sectional area of the prism.

_____ cm² **[2 marks]**

b Calculate the volume of the prism.

_____ cm³ **[2 marks]**

D

2 A cylinder has a radius of 4 cm and a height of 10 cm.

What is the volume of the cylinder?

Give your answer in terms of π.

_____ **[2 marks]**

C

3 A prism has a cross sectional area of 18 cm².

The volume of the prism is 117 cm³.

What is the length of the prism?

_____ **[2 marks]**

C

PS 4 A cylinder has a height equal to its diameter.

The volume is 128π cm³.

What is the radius of the cylinder?

_____ **[3 marks]**

1 Calculate the length of the side marked x in this right-angled triangle.

Give your answer to 1 decimal place.

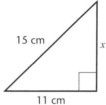

_____ cm **[3 marks]**

2 Calculate the length of the side marked x in this right-angled triangle.

Give your answer to 1 decimal place.

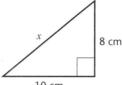

_____ cm **[3 marks]**

3 A flagpole 4 m tall is supported by a wire that is fixed at a point 2.1 m from the base of the pole.

How long is the wire?

(The length is marked x on the diagram.)

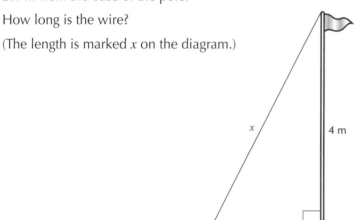

_____ m **[3 marks]**

PS 4 These two triangles have the same height.

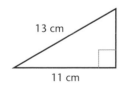 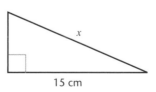

Work out the length x.

Give your answer to 1 decimal place.

_____ **[4 marks]**

Geometry record sheet

Name _____   **Marks** _____

Form _____   **Percentage** _____

Date _____   **Grade** _____

Page number	Question number	Topic	Mark	Comments
155	1	Perimeter and area	/4	
	2	Perimeter and area	/4	
	3	Perimeter and area	/2	
156	1	Perimeter and area	/3	
	2	Perimeter and area	/4	
	3	Perimeter and area	/3	
157	1	Area	/4	
	2	Area	/2	
	3	Area	/2	
158	1	Symmetry	/5	
	2	Symmetry	/2	
	3	Symmetry	/2	
159	1	Angles	/3	
	2	Angles	/4	
	3	Angles	/3	
160	1	Angles	/5	
	2	Angles	/3	
	3	Angles	/3	
161	1	Polygons	/5	
	2	Polygons	/4	
	3	Polygons	/3	
162	1	Parallel lines and angles	/5	
	2	Parallel lines and angles	/8	
	3	Parallel lines and angles	/2	
163	1	Quadrilaterals	/3	
	2	Quadrilaterals	/4	
	3	Quadrilaterals	/4	
164	1	Bearings	/8	
	2	Bearings	/4	
	3	Bearings	/3	
165	1	Circles	/4	
	2	Circles	/4	
	3	Circles	/2	
	4	Circles	/3	
166	1	Scales	/4	
	2	Scales	/2	
	3	Scales	/2	
167	1	Scales and drawing	/5	
	2	Scales and drawing	/5	
	3	Scales and drawing	/2	
168	1	3-D drawing	/2	
	2	3-D drawing	/2	
	3	3-D drawing	/2	
169	1	Congruency and tessellations	/2	
	2	Congruency and tessellations	/2	
	3	Congruency and tessellations	/2	

Page number	Question number	Topic	Mark	Comments
170	1	Transformations	/4	
	2	Transformations	/3	
	3	Transformations	/2	
171	1	Transformations	/5	
	2	Transformations	/4	
	3	Transformations	/3	
172	1	Constructions	/6	
	2	Constructions	/4	
	3	Constructions	/3	
173	1	Constructions and loci	/2	
	2	Constructions and loci	/3	
	3	Constructions and loci	/4	
174	1	Units	/5	
	2	Units	/8	
	3	Units	/3	
175	1	Surface area and volume of 3-D shapes	/3	
	2	Surface area and volume of 3-D shapes	/5	
	3	Surface area and volume of 3-D shapes	/3	
176	1	Prisms	/4	
	2	Prisms	/2	
	3	Prisms	/2	
	4	Prisms	/3	
177	1	Pythagoras' theorem	/3	
	2	Pythagoras' theorem	/3	
	3	Pythagoras' theorem	/3	
	4	Pythagoras' theorem	/4	
Total			**/250**	

Successes

1 _____

2 _____

3 _____

Areas for improvement

1 _____

2 _____

3 _____

1   The MacDonald family are Dad, Mum, Alfie and Bernice.

Alfie is x years old.

a   Bernice is two years younger than Alfie.

Write down an expression for Bernice's age, in terms of x.

_____   **[1 mark]**

b   Dad is twice as old as Alfie.

Write down an expression for Dad's age, in terms of x.

_____   **[1 mark]**

c   Mum is twice Bernice's age.

Write down an expression for Mum's age, in terms of x.

_____   **[1 mark]**

d   Write down and simplify an expression for the total age of the family, in terms of x.

_____   **[2 marks]**

2 a   Draw lines to show which algebraic expressions are equivalent.

One line has been drawn for you.

	$3y$
$3y \times y$	$4y$
$3y + y$	$3y + 3$
$2y + y$	$5y + 2$
$3(y + 1)$	y^2
	$3y^2$

[3 marks]

b   Simplify each of these expressions.

i $q + 4q - 2q$

_____   **[1 mark]**

ii $3p \times 5q$

_____   **[1 mark]**

iii $4x + 3 + 5x - 7$

_____   **[1 mark]**

(AU 3) a and b are different positive whole numbers.

Choose values for a and b so that the formula $7a - 3b$

a   evaluates to a positive odd number   _____   **[1 mark]**

b   evaluates to a prime number   _____   **[1 mark]**

c   evaluates to a positive even number.   _____   **[1 mark]**

Expanding and factorising

1 a Expand $5(x - 3)$. _____ **[1 mark]**

b Expand and simplify $2(x + 1) + 2(3x + 2)$.

_____ **[2 marks]**

c A rectangle has length $2x + 3$ and width $x + 3$.

Write down and simplify an expression for the perimeter, in terms of x.

$x + 3$

$2x + 3$

_____ **[2 marks]**

2 a Multiply out and simplify $3(x - 4) + 2(4x + 1)$.

_____ **[2 marks]**

b Factorise each expression.

i $4x + 6$ **ii** $5x^2 + 2x$

_____ _____ **[1 mark each]**

3 Expand and simplify each expression.

a $2x(3x - 4y) - x(x + 3y)$ _____ **[2 marks]**

b $6(2x - 3y) - 2(x - 3y)$ _____ **[2 marks]**

PS 4 A rectangle with sides 4 and $3x + 1$ has a smaller rectangle with sides 2 and $2x - 1$ cut from it.

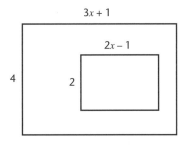

$3x + 1$

$2x - 1$

4

2

Find an expression for the remaining area.

_____ **[2 marks]**

D

1 Solve these equations.

a $\dfrac{x}{3} + 4 = 5$

$x = $ _____ [2 marks]

b $3x + 4 = 1$

$x = $ _____ [2 marks]

c $\dfrac{7x - 2}{3} = 4$

$x = $ _____ [3 marks]

D

2 ABCD is a rectangle.

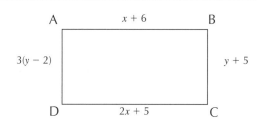

A $x + 6$ B

$3(y - 2)$ $y + 5$

D $2x + 5$ C

a Find the value of x.

$x = $ _____ [3 marks]

b Find the value of y.

$y = $ _____ [3 marks]

D

3 a Solve the equation $2x - 7 = 9$.

$x = $ _____ [2 marks]

b The solution to the equation $2x + 5 = 8$ is $x = 1\frac{1}{2}$.

Zoe thinks that the solution to $2(x + 5) = 8$ is also $x = 1\frac{1}{2}$.

Explain why Zoe is wrong.

_____

_____ [2 marks]

C

PS 4 Here are two flow diagrams.

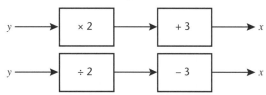

$y \longrightarrow \boxed{\times 2} \longrightarrow \boxed{+ 3} \longrightarrow x$

$y \longrightarrow \boxed{\div 2} \longrightarrow \boxed{- 3} \longrightarrow x$

Both diagrams have the same input, y, and output, x.

What are the values of x and y?

_____ [3 marks]

1 a I think of a number, multiply it by 3 and add 5. The answer is 26.

    i Set up an equation to describe this.

        _____ **[1 mark]**

    ii Solve your equation to find the number.

        _____ **[2 marks]**

b Solve these equations.

    i $4(3y - 2) = 16$

        $y = $ _____ **[3 marks]**

    ii $5x - 2 = x + 10$

        $x = $ _____ **[3 marks]**

2 Solve these equations.

    a $5x - 2 = 3x + 1$

        $x = $ _____ **[3 marks]**

    b $3(x + 4) = x - 5$

        $x = $ _____ **[3 marks]**

    c $5(x - 2) = 2(x + 4)$

        $x = $ _____ **[3 marks]**

3 Solve these equations.

    a $4(x + 3) = x + 3$

        $x = $ _____ **[3 marks]**

    b $5(2x - 1) = 2(x - 3)$

        $x = $ _____ **[3 marks]**

(AU 4) a and b are positive whole numbers.

Find values of a and b to make the solution to this equation $x = 4$.

$a(x + 1) + b(x - 2) = 16$

        _____ **[3 marks]**

E

1 a In the table, a, b, c and d each represent a different number.

The total of each row is shown at the side of the table.

a	a	a	a	16
a	a	b	b	20
a	a	b	c	21
a	b	c	d	25

Find the values of a, b, c and d.

$a =$ _____

$b =$ _____

$c =$ _____

$d =$ _____ **[4 marks]**

b i Write down an expression for the cost of x ice-lollies at 50p each and two choc-ices at 70p each.

_____ **[1 mark]**

ii The total cost of the x lollies and the two ice-lollies is £3.40.

Work out the value of x.

$x =$ _____ **[2 marks]**

C

2 The triangle has sides, given in centimetres, of x, $3x - 1$ and $2x + 5$.

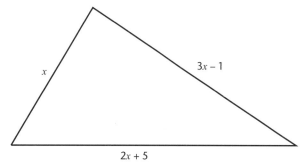

The perimeter of the triangle is 25 cm.

Find the value of x.

$x =$ _____ **[3 marks]**

C

PS 3 Will has eight packets of mints and eight loose mints.

Chas has 11 packets of mints and nine loose mints.

Will says to Chas 'If you give me one of your packets and all of your loose mints we will have the same total number of mints'.

How many mints are in a packet?

_____ **[3 marks]**

1 Use trial and improvement to solve the equation $x^3 + 4x = 203$.

The first two entries of the table are filled in.

Complete the table to find the solution.

Give your answer to 1 decimal place.

Guess	$x^3 + 4x$	Comment
5	145	Too low
6	240	Too high

$x =$ _____ **[3 marks]**

2 Darlene is using trial and improvement to find a solution to

$2x + \dfrac{2}{x} = 8$

The table shows her first trial.

Complete the table to find the solution.

Give your answer to 1 decimal place

Guess	$2x + \dfrac{2}{x}$	Comment
3	6.66	Too low

$x =$ _____ **[4 marks]**

PS 3 A cuboidal juice carton holds 2 litres (2000 cm³). The base is a square. The height is 6 cm more than the base. Use trial and improvement to find the dimensions of the carton.

_____ **[4 marks]**

E

FM 1 A widget weighs x grams.

A whotsit weighs 6 grams more than a widget.

a Write down an expression, in terms of x, for the weight of a whotsit.

_____ grams **[1 mark]**

b Write down an expression, in terms of x, for the total weight of three widgets and one whotsit.

_____ grams **[1 mark]**

c The total weight of three widgets and one whotsit is 27 grams.

Work out the weight of a widget.

_____ grams **[2 marks]**

E

2 a Work out the value of $3p + 2q$ when $p = -2$ and $q = 5$.

_____ **[1 mark]**

b Find the value of $a^2 + b^2$ when $a = 4$ and $b = 6$.

_____ **[1 mark]**

c An aeroplane has f first-class seats and e economy seats.

For a flight, each first-class seat costs £200 and each economy seat costs £50.

i If all seats are taken, write down an expression in terms of f and e for the total cost of all the seats in the aeroplane.

_____ **[1 mark]**

ii If $f = 20$ and $e = 120$, work out the actual cost of all the seats.

_____ **[1 mark]**

C

3 Rearrange each of these formulae to make x the subject.

a $C = \pi x$

$x =$ _____ **[1 mark]**

b $6y = 3x - 9$

Simplify your answer as much as possible.

$x =$ _____ **[2 marks]**

C

PS 4 Keith notices that the cost of six mince pies is £1.24 less than the price of 10 gingerbread men.

Let the price of a mince pie be x pence and the price of a gingerbread man be y pence.

a Express the cost of one gingerbread man, y, in terms of the price of a mince pie, x.

_____ **[2 marks]**

b If the price of mince pie is 21p, how much is an gingerbread man?

_____ **[2 marks]**

Inequalities

1 **a** Solve the inequality: $3x - 4 \leqslant 2$.

_____ **[2 marks]**

b What inequality is shown on the number line?

_____ **[1 mark]**

c Write down all the integers that satisfy both inequalities in parts **a** and **b**.

_____ **[1 mark]**

2 **a** What inequality is shown on the number line?

_____ **[1 mark]**

b Solve these inequalities.

i $\dfrac{x}{2} + 3 > 1$

_____ **[2 marks]**

ii $\dfrac{x + 3}{2} \leqslant 1$

_____ **[2 marks]**

c What integers satisfy both of the inequalities in part **b**?

_____ **[1 mark]**

PS 3 What numbers are being described?

x is an odd number. $2x - 5 \leqslant 14$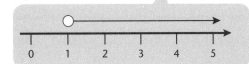

_____ **[2 marks]**

Algebra

F

FM 1 This is a conversion graph between miles and kilometres.

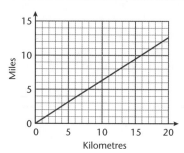

a Approximately how many miles is 5 kilometres? _____ miles **[1 mark]**

b Approximately how many kilometres is 8 miles? _____ miles **[1 mark]**

c Use the graph to work out how many miles is equivalent to 160 kilometres.

_____ miles **[1 mark]**

D

FM 2 Martin does a walk from his house to a viewpoint, 5 kilometres from his house and back again. The distance-time graph shows his journey.

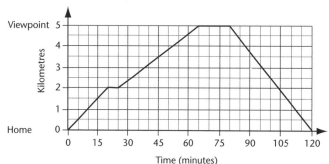

a The viewpoint is uphill from Martin's house.

Martin takes a rest before walking up the steepest part of the hill.

   i How far from home was Martin when he took a rest? _____ km **[1 mark]**

   ii How long did Martin rest? _____ minutes **[1 mark]**

b Martin stopped at the viewpoint before returning home.

He then walked quickly home at a steady pace.

   i How long did it take Martin to return home? _____ minutes **[1 mark]**

   ii What was Martin's average speed on the way home? _____ km/h **[2 marks]**

C

PS 3 A jogger sets off at 10 am from point P to run along a trail at a steady pace of 12 km per hour.

60 minutes later, his wife sets off on a bicycle from P on the same trial at a steady pace of 20 km per hour. After 10 km she gets a puncture which takes 15 minutes to repair. She then sets off again at a steady pace of 20 km per hour.

The couple have left their car at a point 30 km along the trail.

a Represent these journeys on a graph with a horizontal time axis from 10 am to 1 pm and a vertical distance axis from 0 to 35 km.

b Who arrives at the car first and by how long?

_____ **[3 marks]**

1 The table shows some values of the function $y = 3x + 1$ for values of x from –1 to 3.

x	–1	0	1	2	3
y	–2				10

E

a Complete the table of values.   **[1 mark]**

b Use the table to draw the graph of $y = 3x + 1$.

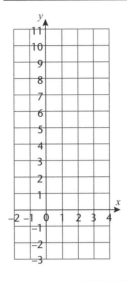

[2 marks]

c What is the x-value when $y = 8$? _____ **[1 mark]**

2 Draw the graph of $y = 2x - 1$ for $-3 \leqslant x \leqslant 3$.

E

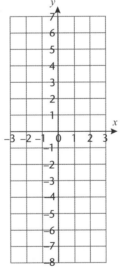

[2 marks]

FM 3 Alf the gas fitter uses this formula to work out how much to charge for a job:

$C = 20 + 30H$

where C is the charge and H is how long the job takes.

Bernice the gas fitter uses this formula:

$C = 30 + 25H$

C

a On the grid draw lines to represent these formulae.

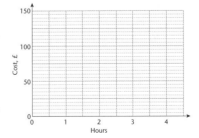

[2 marks]

b An engineer estimates that fitting a boiler will take between 1 and 4 hours.

Which fitter would be the best to employ to do the job?

Give a reason for your answer. _____ **[2 marks]**

Algebra

1 Here are the equations of six lines.

A: $y = 3x + 6$ 　　　　　　　B: $y = 2x - 1$ 　　　　　　C: $y = \dfrac{x}{2} - 1$

D: $y = 3x + 1$ 　　　　　　　E: $y = \dfrac{x}{3} + 1$ 　　　　　　F: $y = 4x + 2$

  a Which line is parallel to line A? 　　　　　　＿＿＿＿＿＿＿＿＿ **[1 mark]**

  b Which line crosses the y-axis at the same point as line B? ＿＿＿＿＿＿ **[1 mark]**

  c Which other two lines intersect on the y-axis? 　　＿＿＿＿＿＿＿＿ **[1 mark]**

  d Write down the gradient of each of these lines.

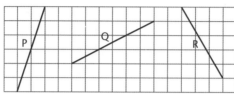

Line P ＿＿＿＿＿＿＿ Line Q ＿＿＿＿＿＿＿ Line R ＿＿＿＿＿＿＿ **[3 marks]**

2 Use the gradient-intercept method to draw
the graph of $y = 3x - 2$ for $-3 \leqslant x \leqslant 3$.

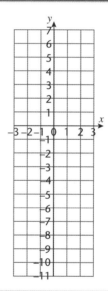

[2 marks]

<u>AU 3</u> The diagram shows a hexagon ABCDEF.

The equation of the line through A and B is $x + y = 4$.

The equation of the line through D and C is $y = -3$.

  a Write down the equation of the lines through the
points

    i E and D

    ii F and E.

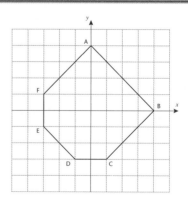

＿＿＿＿＿＿＿＿＿＿ **[2 marks]**

  b The gradient of the line through A and D is 7.

    Write down the gradient of the line through A and C. 　＿＿＿＿＿＿ **[1 mark]**

Quadratic graphs

1 a Complete the table of values for the graph of $y = x^2 + 3$.

x	-3	-2	-1	0	1	2	3
y	12	7					12

[1 mark]

b Draw the graph of $y = x^2 + 3$ for values of x from -3 to 3.

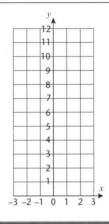

[2 marks]

2 a Complete the table of values for the graph of $y = x^2 - 3x - 4$.

x	-2	-1	0	1	2	3	4
y	6	0			-6		0

[1 mark]

b Draw the graph of $y = x^2 - 3x - 4$ for values of x from -2 to 4.

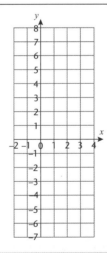

[2 marks]

AU 3 The sketch shows three quadratic graphs, A, B and C.

The equations are $y = x^2 - 4$, $y = x^2$ and $y = x^2 + 3$.

Match each graph with its equation.

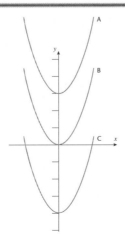

[2 marks]

1 a Complete the table of values for the graph of $y = x^2 - 2x + 1$.

x	−2	−1	0	1	2	3	4
y	9	4	1				

[1 mark]

b Draw the graph of $y = x^2 - 2x + 1$ for values of x from −2 to 4.

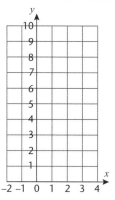

[2 marks]

c Use the graph to find the x-values when $y = 6$. _____ **[1 mark]**

d Use the graph to solve the equation $x^2 - 2x + 1 = 0$. _____ **[1 mark]**

2 a Complete the table of values for the graph of $y = x^2 + 2x - 1$.

x	−4	−3	−2	−1	0	1	2
y		2	−1				7

[1 mark]

b Draw the graph of $y = x^2 + 2x - 1$ for values of x from −4 to 2.

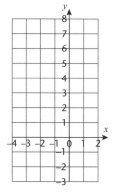

[2 marks]

c Use the graph to find the x-values when $y = 1.5$. _____ **[1 mark]**

d Use the graph to solve the equation $x^2 + 2x - 1 = 0$. _____ **[1 mark]**

AU 3 The first graph shows $y = x^2 - 4x + 3$ and the second shows $y = x^2 - 4x$.

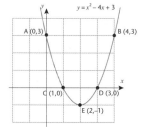 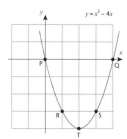

Some of the points on $y = x^2 - 4x + 3$ are shown.

Using these, or otherwise, write down the coordinates of the points P, Q, R, S and T.

_____ **[3 marks]**

Algebra

F

1 **a** Here are three lines of a series of number calculations.

1	=	1	=	1^2
1 + 3	=	4	=	2^2
1 + 3 + 5	=	9	=	3^2
___	=	___	=	___
___	=	___	=	___

   Complete the next two lines of the pattern. **[2 marks]**

b 1, 3, 5, 7, 9, 11, … are the **odd numbers**.

   What is the 50th odd number? _____ **[1 mark]**

c 1, 4, 9, … are the **square numbers**.

   What is the 15th square number? _____ **[1 mark]**

E

2 Squares are used to make patterns.

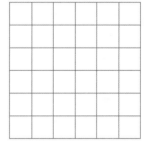

Pattern 1 Pattern 2   Pattern 3   Pattern 4   Pattern 5

a Draw pattern 5. **[1 mark]**

b Complete the table that shows the number of squares in each pattern.

Pattern number	1	2	3	4	5
Number of squares	1	3			

[2 marks]

c Describe, in words, the rule for continuing the number of squares.

   _____

   _____ **[1 mark]**

C

PS 3 When the German mathematician Carl Gauss (1777–1855) was in primary school, his teacher wanted to keep the class quiet for a time so he told them to add up all the whole numbers from 1 to 100. Gauss came up with the answer 5050 in a few seconds.

This is Gauss's method

1 + 2 + 3 + 4 + ……+ 50 + 51 …... + 97 + 98 + 99 + 100

50 × 101 = 5050

Use Gauss's method to work out the sum of all the whole numbers from 1 to 400.

_____ **[3 marks]**

AQA 2 EDEXCEL 1 OCR 1

C

1 The *n*th term of a sequence is $4n + 1$.

a Write down the first three terms of the sequence.

_____ **[1 mark]**

b Which term of the sequence is equal to 29?

_____ **[1 mark]**

c Explain why 84 is not a term in this sequence.

_____

_____ **[1 mark]**

d What is the *n*th term of the sequence 3, 10, 17, 24, 31, ___?

_____ **[2 marks]**

C

2 Matches are used to make patterns with hexagons.

Pattern 1   Pattern 2   Pattern 3   Pattern 4

a Complete the table that shows the number of matches used to make each pattern.

Pattern number	1	2	3	4	5
Number of matches	6	11			

[2 marks]

b How many matches will be needed to make the 20th pattern?

_____ **[1 mark]**

c How many matches will be needed to make the *n*th pattern?

_____ **[2 marks]**

C

(PS 3) Sequence A has an *n*th term $2n + 3$.

Sequence B has an *n*th term $5n - 1$.

a Write down the first 10 terms of each sequence.

_____ **[2 marks]**

b The first term that both sequences have in common is 9.

Find the *n*th term of the sequence formed by all the common terms of the two sequences.

_____ **[2 marks]**

PS 1 R is an odd number, Q is an even number, P is a prime number.

State whether these expressions are *always even*, *always odd* or *could be either*.

		Always even	Always odd	Could be either	
a	$R + Q$	☐	☐	☐	[1 mark]
b	RQ	☐	☐	☐	[1 mark]
c	$P + Q$	☐	☐	☐	[1 mark]
d	R^2	☐	☐	☐	[1 mark]
e	$R + PQ$	☐	☐	☐	[1 mark]

2 a n is a positive integer.

Explain why $2n$ is always an even number.

_____

_____ [1 mark]

C

b Zoe says that when you square an even number you always get a multiple of 4.

Show that Zoe is correct.

_____

_____ [2 marks]

3 Triangles are used to make patterns.

 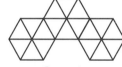 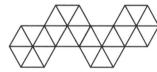

Pattern 1   Pattern 2   Pattern 3

E

a Complete the table that shows the number of triangles used to make each pattern.

Pattern number	1	2	3	4	5
Number of triangles	12				

[2 marks]

b How many triangles will be needed to make the nth pattern?

_____ [2 marks]

PS 4 Two sequences are:

91, 86, 81, 76, 71, ….

1, 5, 9, 13, 17, ….

The nth term of both sequences is the same number.

Which term is this and what is the number?

_____ [3 marks]

C

Algebra record sheet

Name _____ **Marks** _____

Form _____ **Percentage** _____

Date _____ **Grade** _____

Page number	Question number	Topic	Mark	Comments
178	1	Basic algebra	/5	
	2	Basic algebra	/6	
	3	Basic algebra	/3	
179	1	Expanding and factorising	/5	
	2	Expanding and factorising	/4	
	3	Expanding and factorising	/4	
	4	Expanding and factorising	/2	
180	1	Linear equations	/7	
	2	Linear equations	/6	
	3	Linear equations	/4	
	4	Linear equations	/3	
181	1	Linear equations	/9	
	2	Linear equations	/9	
	3	Linear equations	/6	
	4	Linear equations	/3	
182	1	Linear equations	/7	
	2	Linear equations	/3	
	3	Linear equations	/3	
183	1	Trial and improvement	/3	
	2	Trial and improvement	/4	
	3	Trial and improvement	/4	
184	1	Formulae	/4	
	2	Formulae	/4	
	3	Formulae	/3	
	4	Formulae	/4	
185	1	Inequalities	/4	
	2	Inequalities	/6	
	3	Inequalities	/2	
186	1	Graphs	/3	
	2	Graphs	/5	
	3	Graphs	/3	
187	1	Linear graphs	/4	
	2	Linear graphs	/2	
	3	Linear graphs	/4	
188	1	Gradient-intercept	/6	
	2	Gradient-intercept	/2	
	3	Gradient-intercept	/3	
189	1	Quadratic graphs	/3	
	2	Quadratic graphs	/3	
	3	Quadratic graphs	/2	
190	1	Quadratic graphs	/5	
	2	Quadratic graphs	/5	
	3	Quadratic graphs	/3	

Page number	Question number	Topic	Mark	Comments
191	1	Pattern	/4	
	2	Pattern	/4	
	3	Pattern	/3	
192	1	The nth term	/5	
	2	The nth term	/5	
	3	The nth term	/4	
193	1	Sequences	/5	
	2	Sequences	/3	
	3	Sequences	/4	
	4	Sequences	/3	
Total			**/220**	

Successes

1 _____

2 _____

3 _____

Areas for improvement

1 _____

2 _____

3 _____

Area of trapezium $= \frac{1}{2}(a + b)h$

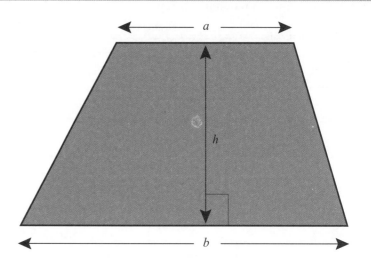

Volume of prism = area of cross-section $\times$ length

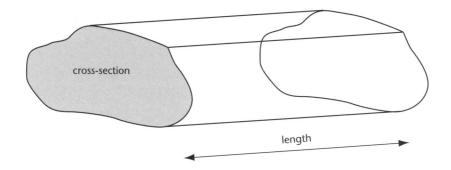

3-D shape A solid shape that has three dimensions (height, width, depth).

addition One of the basic operations of arithmetic. The process of combining two or more values to find their total value. Addition is the inverse operation to subtraction.

algebra Field of mathematics using letters to represent numbers.

allied angles When two parallel lines are cut by a third line (transversal), two pairs of allied angles are formed between the lines, each pair on one side of the transversal. Each pair of allied angles add up to 180°.

alternate angles When two parallel lines are cut by a third line (transversal), two pairs of alternate angles are formed between the lines, one on each side of the transversal. Alternate angles are of equal size.

angle The space (usually measured in degrees [°]) between two intersecting lines or surfaces (planes). The amount of turn needed to move from one line or plane to the other.

angle bisector A straight line or plane that divides an angle in half.

angles in a triangle The three interior (inside) angles of a triangle add up to 180°.

angles around a point The angles around a point add up to 360°.

anticlockwise Turning in the opposite direction to the movement of the hands of a clock. (Opposite of clockwise.)

appropriate Describes an action, data or information which is **relevant** to the situation or investigation.

approximate An inexact value that is accurate enough for the current situation.

arc A curve forming part of the circumference of a circle.

area Measurement of the flat space a shape occupies. Usually measured in square units or hectares. (*See also* surface area.)

average A single number that represents or typifies a collection of values. The three commonly used averages are mode, mean, and median.

average speed A single value of speed that represents the speed taken for a whole journey. The speed of the whole journey if it had been completed at one constant speed.

axis (plural: axes) A fixed line used for reference, along or from which distances or angles are measured.

bar chart A diagram where quantities are represented by rectangles of the same width but different, appropriate heights.

base The bottom of a shape or solid.

bearing The direction relative to a fixed point.

best buy A purchase that gives best value for money spent.

bisect To divide into two equal parts. You can bisect a line or an angle.

bisector A point, line or plane that bisects something. (*See also* perpendicular bisector.)

brackets The symbols '(' and ')' which are used to separate part of an expression. This may be for clarity or to indicate a part to be worked out individually. When a number or value is placed immediately before an expression or value inside a pair of brackets, the two are to be multiplied together. For example, $6a(5b + c) = 30ab + 6ac$.

calculator An electronic device for working out mathematical operations. It is used by pressing keys and the results are shown on the screen.

cancel A fraction can be simplified to an equivalent fraction by dividing the numerator and denominator by a common factor. This is called cancelling.

centilitre (cl) A metric unit of volume or capacity. One hundredth of a litre. 100 cl = 1 litre.

centimetre (cm) A metric unit of length. One hundredth of a metre. 100 cm = 1 m.

centre of rotation The fixed point around which a shape is rotated or turned.

certain Definite. An event is definitely going to occur. The probability that the event will occur = 1.

chance The likelihood, or probability, of an event occurring.

chord A line joining two points on the circumference of a circle.

circle A circle is the path of a point that is always equidistant from another point (the centre).

circumference The outline of a circle. The distance all the way around this outline.

clockwise Turning in the same direction as the movement of the hands of a clock. (Opposite of anticlockwise.)

coefficient The number in front of an unknown quantity (the letter) in an algebraic term. For example, in $8x$, 8 is the coefficient of x.

column A vertical list of numbers or values. The vertical parts of a table. A way of arranging numbers to be added or subtracted.

column method (or traditional method) When adding or subtracting numbers without using a calculator, you can write the numbers in columns by lining up the units digits. This method can also be used for multiplying and dividing by single-digit numbers.

combined events Two or more independent events that occur during a trial.

compasses Also called a pair of compasses, an instrument used for drawing circles and measuring distances.

compass rose A circular diagram showing the main compass directions.

compound Made up from more than one thing. (*See* compound shape.)

compound shape A shape made from two or more simpler shapes. For example, a floor plan could be made from a square and a rectangle joined together.

congruent shapes Shapes having exactly the same size and shape.

consecutive terms Next to each other. For example, Monday and Tuesday are consecutive days, 7 and 8 are consecutive numbers.

consistency A set of data is consistent if all the values are within a small or common range. If a few of the results fall outside this range, the data is not consistent and these 'rogue' results will affect or skew the conclusions drawn from the data.

constant term A term in an algebraic expression that does not change, the number term. For example, in $6x^2 + 5x + 7$, 7 is the constant term.

construct To draw angles, lines or shapes accurately, according to given requirements.

continuous data Data that can be measured rather than counted, such as weight and height.

conversion factor A number that is used to convert a measurement in one unit to another unit. For example, $\times \frac{5}{8}$ converts kilometres to miles.

conversion graph A graph that can be used to convert from one unit to another. It is drawn by joining two or more points where the equivalence is known. Sometimes, but not always, it will pass through the origin.

correlation One measurement is affected by or affects another. For example, weight and height may correlate, but weight and hair colour do not.

corresponding angles When two parallel lines are cut by a third line (transversal), four pairs of corresponding angles are formed: an angle formed by the intersection of the transversal with one of the parallel lines corresponds with the angle in the same position formed by the intersection of the transversal with the other parallel line (think of translation). Corresponding angles are equal.

cross-section The shape of a slice through a solid. Depending on where the cut is made, the cross-section of a cone could be a circle, a triangle, an ellipse or a parabola.

cube 1. A solid with six identical square faces.

cube 2. The result of raising a number to the power of three. For example, 'two cubed' is written: 2^3, which is $2 \times 2 \times 2 = 8$.

cube root The cube root of a number is the value that must be multiplied by itself three times to get the number. The cube root of 27 is 3 (written $\sqrt[3]{27} = 3$), because $3 \times 3 \times 3 = 27$.

cubic centimetre A unit of volume. A cube of side 1 cm has a volume of 1 cm³. $100 \times 100 \times 100$ cm³ = 1 m³.

cubic metre A unit of volume. A cube of side 1 metre has a volume of 1 m³.

cubic millimetre A unit of volume. A cube of side 1 millimetre has a volume of 1 mm³. $10 \times 10 \times 10$ mm³ = 1 cm³.

cuboid A rectangular 3-D (box) shape.

cylinder A solid or hollow prism with circular ends and uniform (unchanging) cross-section. The shape of a can of baked beans or length of drainpipe.

data Information, often of a numerical type, collected during a survey.

data collection sheet A form or table which is used for recording data collected during a survey.

data handling cycle The process of conducting a survey. It consists of four parts: outlining the problem, stating how the data will be collected, stating how the data will be processed, and drawing a conclusion.

decimal Any number using base 10 for the number system. It usually refers to a number written with one or more decimal places.

decimal equivalents A number can be written as decimal, a fraction or a percentage. A fraction or percentage can be converted to a decimal equivalent (a decimal of the same value).

decimal fraction Usually refers to the part of a decimal number after (to the right of) the decimal point, that is, the part less than 1.

decimal place Every digit in a number has a place value (hundreds, tens, ones, etc.). The places after (to the right of) the decimal point have place values of tenths, hundredths, etc. These are called the decimal places.

decimal point The dot used to separate the integer (whole number) place values from the fraction place values (tenths, etc.)

denominator The number under the line in a fraction. It tells you the denomination, name or family of the fraction. For example, a denominator of 3, tells you you are thinking about thirds; the whole thing has been divided into three parts. (*See also* numerator.)

diameter A straight line across a circle, from circumference to circumference and passing through the centre. It is the longest chord of a circle and two radii long. (*See also* radius.)

difference The result of the subtraction of two numbers; the amount by which one number is greater than another number.

digit A number symbol. Our (decimal or denary) number system uses the digits 0, 1, 2, 3, 4, 5, 6, 7, 8 and 9.

direct proportion Two values or measurements may vary in direct proportion. That is, if one increases, then so does the other.

distance The separation (usually along a straight line) of two points.

distance-time graph A graph showing the variation of the distance of an object from a given point during an interval of time. Usually, time is measured along the horizontal axis, and distance is measured up the vertical axis.

division One of the basic operations of arithmetic. Division shows the result of sharing. For example, if you share 12 books among 3 people, they get 4 books each ($12 \div 3 = 4$). It is also used to calculate associated factors. For example, 'How many threes in twelve? ($12 \div 3 = 4$) There are four threes in twelve (or $4 \times 3 = 12$).' It is the inverse operation to multiplication. It can be written using A ÷ B, A/B or $\frac{A}{B}$.

do the same to both sides To keep an equation balanced, you must do the same thing to both sides. If you add something to one side, you must add the same thing to the other side. If you double one side, you must double the other side, etc. If you are manipulating a fraction, you must do the same thing to the numerator and the denominator to keep the value of the fraction unchanged. However you can only multiply or divide the numbers. Adding or subtracting will alter the value of the fraction.

dual bar chart This shows two bar charts on one set of axes. It might show the heights of boys and the heights of girls, rather than the heights of all children.

elevation An elevation is the view of a 3-D shape when it is seen from the front or from another side.

enlargement A transformation of a plane figure or solid object that increases the size of the figure or object by a scale factor but leaves it the same shape.

equally likely Two events are described as equally likely if the probabilities of the occurrence of each of the events are equal. For example, when a die is thrown, the outcomes 6 and 2 are equally likely. (They both have a probability of $\frac{1}{6}$.)

equation A number sentence where one side is equal to the other. An equation always contains an equals sign (=).

equivalent The same, equal in value. For example, equivalent fractions, equivalent expressions.

equivalent fractions Equivalent fractions are fractions which can be cancelled down to the same value, such as $\frac{10}{20} = \frac{5}{10} = \frac{1}{2}$.

estimate (as a verb) To state or calculate a value close to the actual value by using experience to judge a distance, weight, etc. or by rounding numbers to make the calculation easier.

event Something that happens. An event could be the toss of a coin, the throw of a die or a football match.

expand Make bigger. Expanding brackets means you must multiply the terms inside a bracket by the number or letters outside. This will take more room to write, so you have 'expanded' the expression.

expectation Something you expect to happen.

expression Collection of symbols representing a number. These can include numbers, variables (x, y, etc.), operations (+, ×, etc.), functions (squaring, etc.), but there will be no equals sign (=).

exterior angle The exterior angles of a polygon are outside the shape. They are formed when a side is produced (extended). An exterior angle and its adjacent interior angle add up to 180°.

factor A whole number that divides exactly into a given number.

factorisation Finding one or more factors of a given number.

foot (ft) An imperial measurement of length, about 15 cm long. 12 inches = 1 foot, 3 feet = 1 yard.

formula (plural: formulae) An equation that enables you to convert or find a measurement from another known measurement or measurements. For example, the conversion formula from the Fahrenheit scale of temperature to the more common Celsius scale is $\frac{C}{5} = \frac{(F-32)}{9}$ where C is the temperature on the Celsius scale and F is the temperature on the Fahrenheit scale.

fraction A fraction means 'part of something'. To give a fraction a name, such as 'fifths' we divide the whole amount into **equal** parts (in this case five equal parts). A 'proper' fraction represents an amount less than one (the numerator is smaller than the denominator). Any two numbers or expressions can be written as a fraction, i.e. they are written as a numerator and denominator. (*See also* numerator *and* denominator.)

frequency How often something occurs.

frequency table A table showing values (or classes of values) of a variable alongside the number of times each one has occurred.

front elevation The view of a 3-D object when seen from the front.

gallon (gal) An imperial measurement of volume and capacity. A bucket holds about 2 gallons. 8 pints = 1 gallon.

gradient How steep a hill or the line of a graph is. The steeper the slope, the larger the value of the gradient. A horizontal line has a gradient of zero.

gradient-intercept The point at which the gradient of a curve or line crosses an axis.

gram (g) A metric unit of mass. 1000 grams = 1 kilogram.

grouped data Data from a survey that is grouped into classes.

grouped frequency A method of tabulating data by grouping it into classes. The frequency of data values that occur within a class is recorded as the frequency of that class.

height How tall something is. The linear measurement of a shape from top to bottom.

hexagon A polygon with six sides. The sum of all the interior angles of a hexagon is 720°. A regular hexagon has sides of equal length and each of the interior angles is 120°.

highest common factor (HCF) When all the factors of two or more numbers are found, some numbers will have factors in common (the same factors). For example, 6 has factors 1, 2, 3 and 6. 9 has factors 1, 3, and 9. They both have the factors 1 and 3. (1 and 3 are common factors.) The greatest of these is 3, so 3 is the highest common factor.

historical data Data cannot always be found by experiment or from a survey. Data collected by other people, sometimes over a long period of time, is called historical data. Weather records would provide historical data.

hypotenuse The longest side of a right-angled triangle. The side opposite the right angle.

image In geometry the 'image' is the result of a transformation.

imperial The description of measurements used in the UK before metric units were introduced. They often have a long history (for example originating in Roman times) and are commonly based on units of twelve or sixteen rather than ten, as used by the metric system.

impossible If something cannot happen, it is said to be impossible. The probability of it happening = 0.

improper fraction An improper fraction is a fraction whose numerator is greater than the denominator. The fraction could be rewritten as a mixed number. For example, $\frac{7}{2} = 3\frac{1}{2}$.

inch (in) An imperial unit of length. One inch is about $2\frac{1}{2}$ cm long. 12 inches = 1 foot.

index (plural: indices) A power or exponent. For example, in the expression 3^4, 4 is the index, power or exponent.

index form A way of writing numbers (particularly factors) using indices. For example, $2700 = 2 \times 2 \times 3 \times 3 \times 3 \times 5 \times 5$. This is written more efficiently as $2^2 \times 3^3 \times 5^2$.

indices (*See* index.)

inequality An equation shows two numbers or expressions that are equal to each other. An inequality shows two numbers or expressions that are not equal. Instead of the equals (=) sign, the symbol ≠ is used. If it is known which number is the greater or smaller, >, <, ⩾ or ⩽ could be used.

interior angle An angle between the sides inside a polygon. An internal angle. (*See also* allied angles.)

inverse flow diagram A flow diagram to show the reverse operation or series of operations. For example, the inverse of 2 (+ 3) (× 4) → 20 would be 20 (÷ 4) (− 3) → 2.

inverse operations Operations that reverse or cancel out the effect of each other. For example addition is the inverse of subtraction, division is the inverse of multiplication.

isometric grid Dots arranged on paper in a triangular pattern. The pattern makes it easier to draw shapes based on equilateral triangles, parallelograms and trapeziums. It also makes it easier to draw three-dimensional diagrams.

isosceles triangle A triangle with two sides that are equal. It also has two equal angles.

key A key is shown on a pictogram and stem-and-leaf diagram to explain what the symbols and numbers mean. A key may also be found on a dual bar chart, to explain what the bars represent.

kilogram (kg) A metric unit of mass. A bag of sugar has a mass of 1 kg. 1 kilogram = 1000 grams.

kilometre (km) A metric unit of distance. 1 kilometre = 1000 metres.

kite A quadrilateral with two pairs of adjacent sides that are equal. The diagonals on a kite are perpendicular, but only one of them bisects the kite.

length How long something is. We can talk about distances, such as the length of a table, and also time, such as the length of a TV programme.

like terms Terms in algebra that are the same apart from their numerical coefficients. For example: $2ax^2$ and $5ax^2$ are a pair of like terms but $5x^2y$ and $7xy^2$ are not. Like terms can be combined by adding together their numerical coefficients so $2ax^2 + 5ax^2 = 7ax^2$.

likely If an event is likely to occur, there is a good chance that it will occur. There is no fixed number for its probability, but it will be between $\frac{1}{2}$ and 1.

line graph A graph constructed by joining a number of points together.

line of best fit When data from an experiment or survey is plotted on graph paper, the points may not lie in an exact straight line or smooth curve. You can draw a line of best fit by looking at all the points and deciding where the line should go. Ideally, there should be as many point above the line as there are below it.

line of symmetry A mirror line. (*See also* symmetry.)

linear equation An equation of the form $y = ax + b$, where x is limited to the power of 1. The graph of a linear equation is a straight line.

linear graph A straight-line graph from an equation such as $y = 3x + 4$.

linear inequality An inequality usually involving one variable not raised to any power, such as $x < 2 + 4x$. A **strict** inequality will use the symbol < or >, an **inclusive** inequality uses the symbol ⩽ or ⩾.

linear sequence A pattern of numbers where the difference between consecutive terms is always the same.

litre (l) A metric measure of volume or capacity. 1 litre = 1000 millilitres = 1000 cubic centimetres.

loci (*See* locus.)

locus (plural: loci) The locus of a point is the path taken by the point following a rule or rules. For example, the locus of a point that is always the same distance from another point is the shape of a circle.

long division A method of division involving numbers with a large number of digits.

long multiplication A method of multiplication involving numbers with a large number of digits.

lowest common multiple (LCM) Every number has an infinite number of multiples. Two (or more) numbers might have some multiples in common (the same multiples). The smallest of these is called the lowest (or least) common multiple. For example, 6 has multiples of 6, 12, 18, 24, 30, 36, etc. 9 has multiples of 9, 18, 27, 36, 45, etc. They both have multiples of 18 and 36, 18 is the LCM.

mean The mean value of a sample of values is the sum of all the values divided by the number of values in the sample. The mean is often called the average, although there are three different concepts associated with 'average': mean, mode and median.

measurement Finding the size, quantity or amount of an item expressed in appropriate units. In a geometric shape, you can measure lengths, area, volume and angles which are shown in units such as metres, square centimetres, litres and degrees. Linear measure indicates measurement of length; square measure indicates measurement of area and cubic measure indicates measurement of volume.

median The middle value of a sample of data that is arranged in order. For example, the sample 3, 2, 6, 2, 3, 7, 4 may be arranged in order as follows 2, 2, 3, 3, 4, 6, 7. The median is the fourth value, which is 3. If there is an even number of values the median is the mean of the two middle values, for example, 2, 3, 6, 7, 8, 9, has a median of 6.5.

metre (m) A metric unit of length. 1 metre is approximately the arm span of a man. 1 metre = 100 centimetres.

metric A system of units of measurement where the sub-units are related by multiplying or dividing by ten.
For example, for mass,
1 kilogram = 10 hectograms,
1 hectogram = 10 decagrams,
1 decagram = 10 grams,
1 gram = 10 decigrams,
1 decigram = 10 centigrams,
1 centigram = 10 milligrams.
The basic units of length and volume are metres and litres.

middle value The middle value of a sample of data is needed to find the median of that data. The values must be arranged in order first.

mile (m) An imperial unit of length. One mile is almost 2 km.
1 mile = 1760 yards.

millilitre (ml) A metric unit of volume or capacity. One thousandth of a litre. 1000 ml = 1 litre.

millimetre (mm) A metric unit of length. One thousandth of a metre. 1000 mm = 1 metre.

mirror line A line where a shape is reflected exactly on the other side. (*See also* line of symmetry.)

mixed number A number written as a whole number and a fraction. For example, the improper fraction $\frac{5}{2}$ can be written as the mixed number $2\frac{1}{2}$.

modal class If data is arranged in classes, the mode will be a class rather than a specific value. It is called the modal class.

mode The value that occurs most often in a sample. For example, the mode or modal value of the sample: 2, 2, 3, 3, 3, 3, 4, 5, 5, is 3.

multiple The multiples of a number are found by multiplying the number by each of 1, 2, 3, ... For example, the multiples of 10 are 10, 20, 30, 40, etc.

multiplication A basic operation of arithmetic. Multiplication is associated with repeated addition. For example:
$4 \times 8 = 8 + 8 + 8 + 8 = 32$.
Multiplication is the inverse of division.

multiply (*See* multiplication.)

mutually exclusive If the occurrence of a certain event means that another event cannot occur, the two events are mutually exclusive. For example, if you miss a bus you cannot also catch the bus.

negative (in maths) Something less than zero. The opposite of positive. (*See also* negative gradient, negative number *and* positive.)

negative coordinates To plot a point on a graph, two coordinates (x and y) are required. If either the x or y values are on the negative part of the number line, they are negative coordinates.

negative correlation If the effect of increasing one measurement is to decrease another, they are said to show negative correlation. For example, the time taken for a certain journey will have negative correlation with the speed of the vehicle. The slope of the line of best fit has a negative gradient.

negative gradient When a line slopes down from left to right it has a negative gradient. If it slopes up from left to right it has a positive gradient.

negative number Describes a number whose value is less than zero. For example, –2, –4, –7.5, are all negative numbers. (*See also* positive number.)

net If a 3-D shape is made from a sheet of card that can be unglued and folded down flat it will show a net of that solid. Any diagram that could be folded to make a 3-D shape is a net of the shape.

no correlation If the points on a scatter graph are random and do not appear to form a straight line, the two measurements show no correlation. One does not affect the other.

***n*th term** The 'general' term in a sequence. It describes the rule used to get any term.

number line A line where all the numbers (whole numbers and fractions) can be shown by points at distances from zero.

number sequence A pattern of numbers that are related by a common difference.

numerator The number above the line in a fraction. It tells you the number of parts you have. For example, $\frac{3}{5}$ means you have three of the five parts. (*See also* denominator.)

object (in maths) You carry out a transformation on an object to form an image. The object is the original or starting shape, line or point. (*See also* enlargement.)

observation Something that is seen. It can also be the result of a measurement during an experiment or something recorded during a survey.

octagon A polygon that has eight sides. A regular octagon has all its sides of equal length, and each of its interior angles measures 135°.

operation (in maths) an action carried out on two or more numbers. It could be addition, subtraction, multiplication or division.

opposite angles (or vertically opposite angles) When two straight lines cross, four angles are formed. The angles on the opposite side of the point of intersection are equal, so there are two pairs of equal opposite angles.

order The sequence of carrying out arithmetic operations.

order of rotational symmetry A shape or pattern can be rotated about a fixed point. If it has to be turned through a full circle before the picture looks the same as when it started, it has an order of rotation of one. If it looks the same two or three times during that complete rotation it has an order of rotation of two or three.

ordered data Data or results arranged in ascending or descending order.

ounce (oz) An imperial unit of mass. 1 ounce is about 25 grams. 16 ounces = 1 pound.

outcome The result of an event or trial in a probability experiment, such as the score from a throw of a die.

parabola The shape of a graph plotted from an equation such as $5x^2 - 7x + 2 = 0$. The equation will have an 'x^2' term.

parallelogram A four-sided polygon with two pairs of equal and parallel opposite sides.

pattern This could be geometric, where pictures or colours are arranged according to a rule, such as using symmetry. Or it could be numerical where numbers are generated according to an arithmetic rule.

pentagon A polygon that has five sides. A regular pentagon has all its sides of equal length, and each of its interior angles measures 108°.

percentage A number written as a fraction with 100 parts. Instead of writing $\frac{}{100}$, we use the symbol %. So $\frac{50}{100}$ is written as 50%.

percentage decrease If an actual amount decreases, the percentage change will be a percentage decrease.

percentage increase If an actual amount increases, the percentage change will be a percentage increase.

percentage multiplier Percentage expressed as a decimal. For example, 23% would be represented by 0.23.

perimeter The outside edge of a shape. The distance around the edge. The perimeter of a circle is called the circumference.

perpendicular At right angles. Two perpendicular lines are at right angles to each other. A line or plane can also be perpendicular to another plane.

perpendicular bisector A line drawn at right angles (90°) to another line which also divides it into two equal parts.

pictogram A pictorial method of representing data on a graph.

pie chart A chart that represents data as slices of a whole 'pie' or circle. The circle is divided into sections. The number of degrees in the angle at the centre of each section represents the frequency.

pint (pt) An imperial unit of volume or capacity. 1 pint is about half a litre. Milk is usually sold in 1-pint or 2-pint cartons. 8 pints = 1 gallon.

place value The value of a digit depends upon its place or position in the number.

plan A drawing of a room or solid shape as if it is seen from directly overhead.

plane A flat surface.

plane of symmetry The plane about which a 3-D shape reflection occurs. The image of an object in a mirror is symmetrical to the object, and the mirror acts as the plane of symmetry.

polygon A closed shape with three or more straight sides.

polyhedron A solid shape with four or more faces that are polygons. A regular polyhedron has regular polygons as faces. There are only five regular polyhedra (tetrahedron, cube, octahedron, dodecahedron, and icosahedron, with 4, 6, 8, 12 and 20 faces). These regular polyhedra make up the set that is called the Platonic solids.

positive (in maths) Something greater than zero. The opposite of negative (*See also* positive gradient, positive number *and* negative.)

positive correlation If the effect of increasing one measurement is to increase another, they are said to show positive correlation. For example, the time taken for a journey in a certain vehicle will have positive correlation with the distance covered. The slope of the line of best fit has a positive gradient.

positive gradient When a line slopes up from left to right it has a positive gradient. If it slopes down from left to right it has a negative gradient.

positive number Describes a number whose value is greater than zero.

pound (lb) An imperial unit of mass. 1 ounce is about half a kilogram. 1 pound = 16 ounces, 14 pounds = 1 stone.

power The number of times a number or expression is multiplied by itself. The name given to the symbol to indicate this, such as ^{2}. (*See also* index.)

prime factor A factor of a number that is also a prime number. (*See also* prime number.)

prime factor tree A diagram showing all the prime factors of a number. (*See also* prime number.)

prime number A number whose only factors are 1 and itself. 1 is not a prime number. 2 is the only even prime number.

prism A 3-D shape whose ends are identical shapes and the other sides are perpendicular to the ends. A box is a rectangular prism. The volume of a prism is calculated by multiplying the area of the uniform cross-section by the length.

probability The measure of the possibility of an event occurring.

probability scale A line divided at regular intervals. It is usually labelled 'impossible', unlikely, even chance, etc., to show the likelihood of an event occurring. Possible outcomes can then be marked along the scale.

product The result of multiplying two or more numbers or expressions together.

proper fraction A fraction in which the numerator is smaller than the denominator. For example, the fraction $\frac{2}{3}$ is a proper fraction and the fraction $\frac{3}{2}$ is not.

protractor An instrument used for measuring angles.

Pythagoras' theorem The theorem states that the square on the hypotenuse of a right-angled triangle is equal to the sum of the squares on the other two sides.

quadratic equation An equation that includes an x^2 term. For example, $y = 5x^2 - 7x + 2$.

quadratic graph A graph plotted from a quadratic equation.

quadrilateral A polygon that has four sides. The square, rhombus, rectangle, parallelogram, kite, and trapezium are all special kinds of quadrilaterals.

quantity A measurable amount of something which can be written as a number or a number with appropriate units. For example, the capacity of a bottle.

questionnaire A list of questions distributed to people so statistical information can be collected.

radius (plural: radii) The distance from the centre of a circle to its circumference.

random Haphazard. A random number is one chosen without following a rule. Choosing items from a bag without looking means they are chosen at random; every item has an equal chance of being chosen.

range The difference between the largest and smallest data values in a given set of data.

ratio The ratio of A to B is a number found by dividing A by B. It is written as A : B. For example, the ratio of 1 m to 1 cm is written as 1 m : 1 cm = 100 : 1. Notice that the two quantities must both be in the same units if they are to be compared in this way.

rational number A rational number is a number that can be written as a fraction, for example, $\frac{1}{4}$ or $\frac{10}{3}$.

raw data Data in the form it was collected. It hasn't been ordered or arranged in any way.

ray method A method of constructing enlargements using lines (that look like rays) drawn from a single point.

rearrangement To change the arrangement of something. An equation can be rearranged using the rules of algebra to help you solve it. Data can be rearranged to help you analyse it.

rectangle A quadrilateral in which all the interior angles are 90°. The opposite sides are of equal length. A rectangle has two lines of symmetry and a rotational symmetry of order 2. The diagonals of a rectangle bisect each other and also bisect the rectangle itself.

recurring decimal A decimal number that repeats forever with a repeating pattern of digits.

reflection The image formed after being reflected. The process of reflecting an object.

regular polygon A polygon that has sides of equal length and angles of equal size.

relative frequency Also known as experimental probability. It is the ratio of the number of successful events to the number of trials. It is an estimate for the theoretical probability.

remainder When a number is divided by a divisor, the result is a quotient and a remainder. The remainder could be zero. For example, 17 divided by 5 gives a quotient of 3 with a remainder of 2. 12 divided by 3 gives a quotient of 4 with no remainder.

representative A number or quantity that is typical of a set of data. A person that is typical of a given group. An average is representative of all the given values.

rhombus A parallelogram that has sides of equal length. A rhombus has two lines of symmetry and a rotational symmetry of order 2. The diagonals of a rhombus bisect each other at right angles and they bisect the figure.

right-angled triangle A triangle with one angle equal to 90°.

rotation Turning. A geometrical transformation in which every point on a figure is rotated through the same angle.

rotational symmetry A shape which can be turned about a point so that it coincides exactly with its original position at least twice in a complete rotation has rotational symmetry.

rounding An approximation for a number that is accurate enough for some specific purpose. The rounded number may be used to make arithmetic easier and is always less precise than the unrounded number.

sample The part of a population that is considered for statistical analysis. The act of taking a sample within a population is called sampling. There are two factors that need to be considered when sampling from a population: 1. The size of the sample. The sample must be large enough for the results of a statistical analysis to have any significance. 2. The way in which the sampling is done. The sample should be representative of the population.

sample space diagram A diagram or table showing all the possible outcomes of events such as throwing dice or tossing coins.

scale drawing An accurate drawing where lengths are reduced from the real-life lengths to ones that can be drawn on paper. The reduction is by a given ratio each time.

scale factor The ratio by which a length or other measurement is increased or decreased.

scalene triangle A triangle where all the sides are of unequal length.

scales 1. A scale on a diagram shows the scale factor used to make the drawing.

scales 2. An instrument used to find the mass or weight of something.

scatter diagram A diagram of points plotted of pairs of values of two types of data. The points may fall randomly or they may show some kind of correlation.

sector A region of a circle, like a slice of a pie, bounded by an arc and two radii.

segment A part of a circle between a chord and the circumference.

sequence A pattern of numbers which are related by a common difference.

shape Could be a 2-D or 3-D shape. Any drawing or object. In mathematics we usually study simple shapes such as squares, prisms, etc. or compound shapes that can be formed by combining two or three simple shapes, such as an ice-cream cone made by joining a hemisphere to a cone.

side 1. A straight line forming part of the perimeter of a polygon. For example, a triangle has three sides.

side 2. A face (usually a vertical face) of a 3-D object, such as the side of a box.

side elevation The view of a 3-D object when seen from a side.

sign A symbol used to represent something such as $\times$, $\div$, = and $\sqrt{}$. The 'sign of a number' means whether it is a positive or negative number.

significant figure The significance of a particular digit in a number is concerned with its relative size in the number. The first (or most) significant figure is the left-most, non-zero digit; its size and place value tell you the approximate value of the complete number. The least significant figure is the right-most digit; it tells you a small detail about the complete number. For example, if we write 78.09 to 3 significant figures we would use the rules of rounding and write 78.1.

simplest form A fraction cancelled down so it cannot be simplified any further. An expression where the arithmetic is completed so that it cannot be simplified any further.

simplify To make an equation or expression easier to work with or understand by combining like terms or cancelling down. For example: $4a - 2a + 5b + 2b = 2a + 7b$, $\frac{12}{18} = \frac{2}{3}$.

solution The result of solving a mathematical problem. Solutions are often given in equation form.

solve Finding the value or values of a variable (x) which satisfy the given equation.

speed The rate at which something moves or happens.

spread (*See* range.)

square 1. A polygon with four equal sides and all the interior angles equal to 90°.

square 2. The result of multiplying a number by itself. For example, 5^2 or 5 squared is equal to $5 \times 5 = 25$.

square numbers Numbers obtained by squaring the whole numbers. For example, 1, 4, 9, 16, 25, are square numbers.

square root The square root of a number is the value that must be multiplied by itself to get the number. The square root of 9 is 3 (written $\sqrt{9} = 3$), because $3 \times 3 = 9$.

stem-and-leaf diagram A way of making a table of statistical data so that the classes are visualised as the stem of a plant and the data values as its leaves.

stone (st) An imperial unit of mass. 1 stone is about 6 kilograms. 1 stone = 14 pounds, 160 stone = 1 ton.

substitution When a letter in an equation, expression or formula is replaced by a number, we have substituted the number for the letter. For example, if $a = b + 2x$, and we know $b = 9$ and $x = 6$, we can write $a = 9 + 2 \times 6$. So $a = 9 + 12 = 21$.

subtraction One of the basic operations of arithmetic. It finds the difference between two numbers. Subtraction is the inverse operation to addition.

sum The result of adding two or more numbers or quantities.

surface area The area of the surface of a 3-D shape, such as a prism. The area of a net will be the same as the surface area of the shape.

survey A questionnaire or interview held to find data for statistical analysis.

symbol A written mark that has a meaning. All digits are symbols of the numbers they represent. $+$, $<$, $\sqrt{}$ and $°$ are other mathematical symbols. A pictogram uses symbols to represent amounts.

symmetry A figure is said 'to have symmetry' if it remains unchanged under a transformation. For example, the letter T has one line of symmetry (a mirror down the middle would produce an identical reflection), the letter N has rotational symmetry of order two (a rotation of 180° would produce an image that looks like an 'N').

tally chart A chart with marks made to record each object or event in a certain category or class. The marks are usually grouped in fives to make counting the total easier.

term 1. A part of an expression, equation or formula. Terms are separated by $+$ and $-$ signs.

term 2. Each number in a sequence or arrangement in a pattern.

terminating Ending. A terminating decimal can be written down exactly. $\frac{33}{100}$ can be written as 0.33, but $\frac{3}{9}$ is 0.3333..... with the 3s recurring forever.

tessellation A shape is said to tessellate if, when its image is translated and/or reflected and/or rotated, the shapes completely fill a space, leaving no gaps. A space filled in this way is said to form a tessellation.

three-figure bearing The angle of a bearing is given with three digits. If the angle is less than 100°, a zero (or zeros) is placed in front, such as 045° for north-east.

time How long something takes. Time is measured in days, hours, seconds, etc.

times tables Tables used to list the multiplication of numbers, often up to 10×10, sometimes up to 12×12.

ton (ton) An imperial unit of mass. 1 ton is about 1 tonne. 1 ton = 160 stone.

tonne (T) A metric unit of mass. 1 tonne is about 1 ton. 1 tonne = 1000 kilograms.

top-heavy A fraction where the numerator is greater than the denominator could be described as top-heavy.

transformation An action such as translation, reflection or rotation.

translation A transformation in which all points of a plane figure are moved by the same amount and in the same direction.

transversal A line that crosses two parallel lines.

trapezium A quadrilateral with one pair of parallel sides.

travel graph (*See* distance-time graph.)

trends A collection of data can be analysed (for example by drawing a time graph) so that any trend or pattern, such as falling profits, may be discovered.

trial An experiment to discover an approximation to the probability of the outcome of an event will consist of many trials where the event takes place and the outcome is recorded.

trial and improvement A method of solving an equation by guessing a solution, then making further educated guesses based on the outcome of the previous one, thus improving each time to get ever closer to the solution.

triangle A three-sided polygon. The interior angles add up to 180°. Triangles may be classified as:
1. scalene: no sides of the triangle are equal in length (and no angles are equal).
2. equilateral: all the sides of the triangles are equal in length (and all the angles are equal).
3. isosceles: two of the sides of the triangle are equal in length (and two angles are equal).
4. A right-angled triangle has an interior angle equal to 90°.

two-way table This links two variables. One is listed in the column headers and the other in the row headers; the combination of the variables is shown in the body of the table.

unit cost The cost of one unit of a commodity, such as the cost per kilogram.

unitary method A method of calculation where the value for one item is found before finding the value for several items.

units 1. Ones, as in hundreds, tens and units.

units 2. A standard measure of a quantity e.g. mass, length, time. For example, the kilogram is a unit of mass.

unlikely An outcome with a low chance of occurring.

unordered data (*See* raw data.)

variable A quantity that can have many values. These values may be discrete or continuous. They are often represented by x and y in an expression.

vector A quantity with magnitude **and** direction.

vertex The points at which the sides of a polygon or the edges of a polyhedron meet.

volume The amount of space occupied by a substance or object or enclosed within a container.

weight How heavy something is. An object's weight is measured on scales or a balance.

width How wide something is. The linear measurement of a shape from side to side.

x-**value** The value along the horizontal axis on a graph.

yard (yd) An imperial unit of length. 1 yard is about 1 metre. 3 feet = 1 yard.

y-**value** The value along the vertical axis on a graph.

π (pronounced 'pi') The numerical value of the ratio of the circumference of a circle to its diameter (approximately 3.14159).

Index

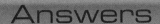

In this mark scheme there are five types of mark.

B marks: These are for questions which are worth only 1 mark, or where there is no method needed.

M marks: These are for correct methods that would lead to the answer.

A marks: These are accuracy marks for correct answers.

E marks: These are for explanations.

Q marks: These are for the quality of written communication.

You will also see other abbreviations:

ft   This means 'follow through' meaning that your answer can be marked correct if you have used an earlier incorrect answer but done all the subsequent working correctly.

112   Statistical representation

Q	Answer	Mark	Comment
1a	13, 6, 2	B1	You will get 1 mark for a partial explanation
1a	ＨＨＴ ///	B1	
1c	50	B1	
1d	Because 21 cars had 1 passenger (21); 13 cars had 2 passengers (26); 8 cars had 3 passengers (24); 6 cars had 4 passengers (24); 2 cars had 5 passengers (10) and 21 + 26 + 24 + 24 + 10 (= 105)	Q2	You will get 1 mark for a partial explanation
2a	95; 70	B1	
2b	⊠ ⊠ and ⊠ ⊠	B1	
2c	305	M1	
3	A: Look at past historical data; B: Test by experiment; C: Find out how many tickets were sold (equally likely outcomes); D: Do a survey	B3	There is no need to write lots of words. A brief answer such as 'Experiment' will often do. You lose a mark for every one you get wrong

113   Statistical representation

Q	Answer	Mark	Comment
1a		B2	You will lose a mark for every mistake
1b	7	B1	
1c	No, student B and student F spent more time doing homework than watching TV	M1	
2a	16 °C	B1	
2b	10am–11am	M1	
2c	11am	M1	
2d	25 °C	M1	
2e	No, anything could happen in four hours, it is too far from the end of the graph	M1	
3	1 division = 2 shirts	B1	Compare the difference in the scales to 14
	20 + 14 + 14 + 10	M1	
	58	A1	

114 Averages

Q	Answer	Mark	Comment
1a	50	B1	
1b	1	B1	
1c	2	B1	
1d	54%	B1	
2ai	$640 \div 10$	M1	
	64	A1	
2aii	25	B1	
2bi	924 kg	B1	
2bii	$14 \times 87 - 924$	M1	
	$294 \div 3$	M1	
	98 kg	A1	
3	5, 8, 9, 9, 9	B3	There may be other answers but you will get part marks for meeting the conditions in the question. e.g. a total of 40, a mode of 9, a range of 4 each get 1 mark

115 Averages and range

Q	Answer	Mark	Comment
1a	24	B1	
1b	2	B1	
1c	2	B1	
1d	$42 \div 24$	M1	
	1.75	A1	
1e	420	B1	
	No as $420 > 400$	Q1	Make sure you use numbers in your conclusion
2ai	46	B1	
2aii	98	B1	
2aiii	$1060 \div 10$	M1	
	106	A1	
2bi	Only 2 people and the lowest value	B1	
2bii	Affected by one extreme value	B1	
2c	No; mode does not tell you anyting about the total score	B1	
3a	$7 \times 9 - 6 \times 8$	M1	
	15	A1	
3b	Facts B and C	B1	

116 Arranging data

Q	Answer	Mark	Comment
1a	2	B1	
1b	2	B1	
1c	Total = 180	M1	
	1.8	A1	
2a	Midpoints 45, 55 etc.	B1	
	$45 \times 27 + 55 \times 39... + 95 \times 12$	M1	
	$13\,000 \div 200$	M1	
	65	A1	

2b		B2	You will lose a mark for every wrong plot
2c	Girls did better; polygons are about the same shape and girls are about 10 marks better	B1	
3	Mid points 62.5, 67.5 etc	B1	
	$62.5 \times 12 + 67.5 \times 15 + \ldots + 82.5 \times 6$	M1	
	$3525 \div 50$	M1	
	70.5	A1	

117 Arranging data

Q	Answer	Mark	Comment
1a	0	7 8 9 9 1	1 2 3 3 3 3 5 6 8 9 2	1 2 2 3 4 4	B2	You will lose a mark for every error
1bi	130	B1	
1bii	140	B1	
1biii	170	B1	
2a	30	B1	
2b	288	B1	You need to include the students whole were not late
	$288 \div 30$	M1	
	9.6	A1	
3a	No as there are three values in the current mode and no current value has two values	B1	
3b	Yes as it will be a whole number now, either 41 or 42	B1	
3c	Cannot tell. The new number may be the same as the old mean, which is 41	B1	

118 Probability

Q	Answer	Mark	Comment
1ai	Impossible	B1	
1aii	Very unlikely	B1	
1aiii	Even	B1	
1aiv	Very likely/certain	B1	
1b		B4	
2ai	$\frac{12}{20} = \frac{3}{5}$	B1	
2aii	0	B1	
2aiii	$\frac{8}{20} = \frac{2}{5}$	B1	
2b	4	B1	
3a	$\frac{2}{5}$	B1	

Workbook answers

3b	2 : 8	M1	
	10	A1	
4	The probability is $\frac{4}{9}$	B1	

119 Using probability

Q	Answer	Mark	Comment
1a	$\frac{47}{50}$	B1	
1b	30	B1	
2a	Mrs Rogers; she has more tickets	B1	
2b	$\frac{5}{900} = \frac{1}{180}$	B1	
2c	$\frac{4}{900} = \frac{1}{225}$	B1	
2d	$\frac{19}{900}$	B1	
3a	Any frequency ÷ 120	M1	
	0.15, 0.06, 0.18, 0.18, 0.29, 0.14	A1	
3b	5; as this had a much higher relative frequency	B1	
4	Any combination of two bags and the probability of white calculated. A and B $\frac{6}{13}$ = 0.4615. A and C $\frac{7}{15}$ = 0.4666. B and C $\frac{7}{16}$ = 0.4375	M1	Show clearly that you are combining two bags. Write 'A and B give 6 white in a total of 13', for example. You will have to use a calculator to work out the fractions as dcimals.
	Two combinations worked out	A1	
	A and C stated and all combinations worked out	A1	

120 Combined events

Q	Answer	Mark	Comment
1a	Sausage, hash browns and beans; sausage, hash browns and toast; bacon, eggs and beans; bacon, eggs and toast; bacon, hash browns and beans; bacon, hash browns and toast	B2	You will lose a mark for any missing values but repeats will not be penalised
1b	$\frac{2}{8} = \frac{1}{4}$	B1	
2a	12	B1	
2bi	$\frac{3}{12} = \frac{1}{4}$	B1	
2bii	$\frac{2}{12} = \frac{1}{6}$	B1	
2ci	Top row: 2, 4, 6, 8, 10, 12; bottom row: 0, 1, 2, 3, 4, 5	B2	You will lose 1 mark for each error
2cii	$\frac{3}{12} = \frac{1}{4}$	B1	
3	P(5 with 1 dice) = $\frac{1}{6}$	B1	
	(2, 3), (4, 1), (1, 4), (3, 2)	M1	
	P(5 with two dice) = $\frac{4}{36} = \frac{1}{9}$	A1	
	1 dice as $\frac{1}{6} > \frac{1}{9}$	Q1	Make sure your conclusion is correct

121 Expectation and two-way tables

Q	Answer	Mark	Comment
1a	As 2 + 3 = 5; the probability of red is $\frac{2}{5}$	B1	
1b	12	B1	
1c	80	B1	

2a	22	B1	
2b	PE	B1	
2c	Maths $\frac{5}{12} \approx 42\%$; Science $\frac{7}{18} \approx 39\%$	B1	
2d	$\frac{12}{40} = \frac{3}{10}$	B1	
3	2, 5, 7 8, 10, 18 10, 15, 25	B3	Check that your final table satisfies the initial conditions. You will lose a mark for every condition that is not met

122 Pie charts

Q	Answer	Mark	Comment
1a	Any frequency × 6 ÷ 360	M1	
	132°, 108°, 72°, 48°	A1	
1b		B2	1 mark for accurate drawing and 1 mark for labelling
1c	160	B1	
2a	80 does not divide by 15 to give a factor of 360	B1	
2b	60, 75, 120	B1	

123 Scatter diagrams

Q	Answer	Mark	Comment
1a		B2	You will lose a mark for any misplots.
1bi	G	B1	
1bii	I	B1	
1c	See graph	B1	
1d	54 minutes	B1	
3	A and Y; B and W; C and Z; D and X	B2	

124 Surveys

Q	Answer	Mark	Comment
1a	<table><tr><td></td><td>Boys</td><td>Girls</td></tr><tr><td>0 < time (hours) ⩽ 4</td><td></td><td></td></tr><tr><td>4 < time (hours) ⩽ 6</td><td></td><td></td></tr><tr><td>6 < time (hours) ⩽ 8</td><td></td><td></td></tr><tr><td>8 < time (hours) ⩽ 10</td><td></td><td></td></tr><tr><td>More than 10 hours</td><td></td><td></td></tr></table>	B2	You will lose a mark for any overlapping responses
1b	Not really; the difference is not that large and the sample of girls was too small	B1	
2a	Leading question	B1	
	Not enough responses	B1	
2b	0p–99p, ☐; £1–£1.99, ☐; £2–£2.99 ☐	B2	You will lose a mark if your responses overlap, if you miss any values, or if you have too wide a range (such as up to £20).

Workbook answers

3	Total for 2009 = 39 + 45 + ... + 45 + 43	M1	
	432 ÷ 12 = 36	A1	
	Range for 2009 = 38	B1	
	The hypothesis is not supported as the means are very close (35 and 36) and so are the ranges (38 and 40)	Q2	Q1 for a partial explanation; not mentioning values, for example

126 Basic number

Q	Answer	Mark	Comment
1ai	72	B1	
1aii	280	B1	
1aiii	80	B1	
1bi	$4 \times 7 = 28$	B1	
1bii	$28 \div 7 = 4$ or $28 \div 4 = 7$	B1	
2a	4	B1	
2bi	$(2 + 4^2) \div 8 = 2.25$	B1	
2bii	$(2 + 4)^2 \div 8 = 4.5$	B1	
3ai	26 (7 + 19)	B1	
3aii	16 (6 + 10)	B1	
3bi	25 (18 + 7)	B1	
3bii	14 (12 + 2)	B1	

127 Basic number

Q	Answer	Mark	Comment
1a	3659	B1	
1b	6864	B1	
1c	1770	B1	
2ai	Ninety-six thousand, nine hundred and twenty-four	B1	
2aii	96 900	B1	
2aiii	97 000	B1	
2bi	9 thousand	B1	
2bii	68 950	B1	
2biii	69 049	B1	
2biv	23 000	B1	
3a	898	B1	
3b	3178	B1	
3c	1225	B1	
4a	1139	B1	
4b	2317	B1	
5	Testing a value for my age	M1	This problem is best attempted using trial and improvement. Guess a value, halve it for my daughter's age and then subtract 14 off each to see if one is 4 times the other
	Testing an improved value	M1	
	42	A1	

128 Basic number

Q	Answer	Mark	Comment
1a	304	B1	
1b	26	B1	
1c	378	B1	

1d	48	B1	
2ai	272p or £2.72	B1	
2aii	228p or £2.28	B1	
2b	24 × 30	M1	
	720	A1	
2c	196 ÷ 7	M1	
	28	A1	
3a	20 − 12.85 = £7.15	B1	
3b	14.55 ÷ 3	M1	
	£4.85	A1	
3c	12 × 9	M1	
	= 108	A1	
4a	Both round to 4 to 1sf	B1	There are other answers. Any valid property is OK
4b	Both have 3 decimal places	B1	
4c	0.365 > 0.363	B1	

129 Fractions

Q	Answer	Mark	Comment
1a	$\frac{9}{12}$ and $\frac{60}{80}$	B1	
1b	Any 18 squares shaded	B1	
1ci	$\frac{14}{20} = \frac{7}{10}$	B1	
1cii	$\frac{6}{20} = \frac{3}{10}$	B1	
2ai	Any 2 squares shaded	B1	
2aii	Any 3 squares shaded	B1	
2b	$\frac{5}{9}$	B1	
3a	The fraction is $\frac{2}{5}$	B1	
3b	$\frac{4}{10}$ or $\frac{6}{15}$ or $\frac{20}{50}$ or $\frac{8}{20}$ etc	B1	
4	$\frac{3}{2}$ is top heavy	B1	
	$\frac{2}{3}$ has odd denominator	B1	
	$\frac{2}{8}$ can be cancelled	B1	

130 Fractions

Q	Answer	Mark	Comment
1a	Any 2 squares shaded. Any 4 squares shaded. Any 6 squares shaded. Any 10 squares shaded	B2	You will get 1 mark for three correct
1bi	12	B1	
1bii	32	B1	
1biii	6	B1	
1biv	2	B1	
1ci	$\frac{4}{7}$	B1	
1cii	$\frac{3}{5}$	B1	
1ciii	$\frac{2}{5}$	B1	
1civ	$\frac{3}{4}$	B1	

Workbook answers

1d	$\frac{13}{20}$; $\frac{7}{10}$; $\frac{4}{5}$	B1	
2ai	$1\frac{4}{5}$	B1	
2aii	$2\frac{3}{7}$	B1	
2aiii	$2\frac{5}{8}$	B1	
2aiv	$7\frac{3}{4}$	B1	
2bi	$\frac{17}{11}$	B1	
2bii	$\frac{11}{8}$	B1	
2biii	$\frac{7}{3}$	B1	
2biv	$\frac{23}{5}$	B1	
3	Outer shaded squares are $\frac{1}{4}$	B1	
	Inner are $\frac{1}{8}$	B1	
	Total $\frac{3}{8}$	B1	

131   Fractions

Q	Answer	Mark	Comment
1ai	6	B1	
1aii	10	B1	
1aiii	2	B1	
1aiv	9	B1	
1bi	$\frac{7}{20}$	B1	
1bii	$\frac{13}{16}$	B1	
1biii	$\frac{1}{2}$	B1	
1biv	$\frac{5}{6}$	B1	
1ci	$\frac{11}{20}$	B1	
1cii	$\frac{5}{16}$	B1	
1ciii	$\frac{1}{2}$	B1	
1civ	$\frac{1}{6}$	B1	
2ai	$\frac{3}{5}$	B1	
2aii	$\frac{2}{5} \times 245$	M1	
	98	A1	
2b	$90 \times 2 + 30$	M1	
	210	A1	
2c	$220 - 55$	M1	
	165	A1	
3	$\frac{1}{4} \times \frac{4}{5} = \frac{1}{5}$	B1	
	$\frac{1}{3} \times \frac{3}{5} = \frac{1}{5}$	B1	
	They all get $\frac{1}{5}$	B1	

132   Fractions

Q	Answer	Mark	Comment
1ai	$\frac{4}{11}$	B1	
1aii	$\frac{1}{6}$	B1	
1aiii	$\frac{1}{8}$	B1	
1aiv	$\frac{3}{4}$	B1	
1bi	$1\frac{1}{2}$	B1	
1bii	$1\frac{1}{2}$	B1	
1biii	2	B1	
1biv	6	B1	
2a	$\frac{1}{6}$	B2	You will get a mark for writing $\frac{250}{1500}$
2b	$\frac{1}{3}$	B2	You will get a mark for writing $\frac{40}{120}$
3a	$\frac{3}{20} \times 18000$	M1	
	£2700	A1	
3b	$\frac{1}{6} \times 1.2$	M1	
	1.4 kg	A1	
4	600 ÷ 120 or 500 ÷ 96	M1	
	5 g/p and 5.2 g/p	A1	
	Offer B as 5.2 > 5	Q1	Make sure you give a clear conclusion using numbers

133   Rational numbers

Q	Answer	Mark	Comment
1ai	0.175	B1	
1aii	0.7$\dot{3}$	B1	
1aiii	0.8$\dot{3}$	B1	
1aiv	0.18	B1	
1bi	0.4444…	B1	
1bii	0.5555…	B1	
1ci	0.2727…	B1	
1cii	0.5454…	B1	
2a	$\frac{6}{11}, \frac{11}{20}, \frac{5}{9}, \frac{14}{25}$	B2	You will lose a mark if any are out of order
2b	$\frac{3}{11}, \frac{1}{4}, \frac{6}{25}, \frac{2}{9}$	B2	You will lose a mark if any are out of order
3a	0.8, 0.4, 0.2, 0.1	B2	You will lose a mark if any wrong answers
3b	1 ÷ 40 = 0.025 as the terms are dividing by 2	B1	

134   Negative numbers

Q	Answer	Mark	Comment
1a	Aberdeen	B1	
1b	9°	B1	
1c	Aberdeen and Bristol	B1	
1d	London	B1	

Workbook answers

2a		B2	You will lose a mark for each wrong answer
2b	−2.2; −2.1; −2.0	B1	
2ci	2	B1	
2cii	−9.5	B1	
3	−3 and −6	B2	One mark for each value

135 Negative numbers

Q	Answer	Mark	Comment
1a	−1	B1	
1b	+6; −6	B1	
1ci	+8	B1	
1cii	−4	B1	
1di	−7	B1	
1dii	11	B1	
2	−11, +2, −2, −1	B2	You will get 1 mark for three correct
3a	Any correct pair such as 1 − 6	B1	
3b	Any correct pair such as − 9 − − 4	B1	

136 More about number

Q	Answer	Mark	Comment
1a	6 and 12	B1	
1b	6 and 10	B1	
1c	<table><tr><td>12</td><td>7</td><td>8</td></tr><tr><td>5</td><td>9</td><td>13</td></tr><tr><td>10</td><td>11</td><td>6</td></tr></table>	B2	You will get 1 mark for three or four correct
2ai	{1, 3, 11, 33}	B1	
2aii	{1, 2, 3, 6, 9, 18}	B1	
2bi	84 or 90	B1	
2bii	85 or 90	B1	
4	60 seconds	B1	
3	3	B1	

137 Primes and squares

Q	Answer	Mark	Comment
1a	11 and 13	B1	
1b	2 is even and prime	B1	
1c	9 and 16	B1	
1d	$5^2 = 25$; $10^2 = 100$	B1	
1e	11 and 13	B2	
2a	Either odd or even	B1	
2b	Either odd or even	B1	
2c	Always odd	B1	
2d	Either odd or even	B1	
3	20 is not square	B1	
	25 only odd	B1	
	36 not multiple of 5	B1	

138	**Roots and powers**		
Q	**Answer**	**Mark**	**Comment**
1a	$\sqrt{49} = 7$ (or $\sqrt{64} = 8$)	**B1**	
1b	$\sqrt{64} = 8$ (or $\sqrt{49} = 7$)	**B1**	
1c	$2^4 = 16 > \sqrt{144} = 12$	**B1**	
1di	13	**B1**	
1dii	125	**B1**	
1ei	4	**B1**	
1eii	256	**B1**	
2a	64, 4; 256, 6; 1024, 4	**B2**	You will lose 1 mark for every wrong answer
2b	4, all odd powers end in 4	**B1**	
2c	$5^6 = 15\,625 > 6^5 = 7776$	**B1**	
3	6, 8 and 5	**B1**	Any two answers for 1 mark
	$\sqrt{25}$, $\sqrt[3]{216}$, 2^3	**B1**	

139	**Powers of 10**		
Q	**Answer**	**Mark**	**Comment**
1a	4 tenths; 0.4 or $\frac{4}{10}$	**B1**	
1b	10^4	**B1**	
1c	10 000 000	**B1**	
1d	100; $\frac{1}{100}$; 10^1; 10^0; 10^{-3}	**B1**	
1ei	1	**B1**	
1eii	5	**B1**	
2ai	370	**B1**	
2aii	250	**B1**	
2bi	0.76	**B1**	
2bii	0.0065	**B1**	
2ci	12 000 000	**B1**	
2cii	360 000	**B1**	
2di	3000	**B1**	
2dii	500	**B1**	
3	1 m = 100 cm	**B1**	
	1 000 000	**B1**	

140	**Prime factors**		
Q	**Answer**	**Mark**	**Comment**
1a	90	**B1**	
1b	$2 \times 5 \times 7$	**B1**	
1c	$2^4 \times 3$	**B1**	
1di	9, 3, 3, 10, 2, 5, 2, 5	**B2**	You will lose 1 mark for each error
1dii	$2^2 \times 3^2 \times 5^2$	**B1**	
2ai	5	**B1**	
2aii	$2 \times 3 \times 5^2$	**B1**	
2bi	3	**B1**	
2bii	$2^3 \times 3^3$	**B1**	

Workbook answers

3	Dividing 2009 by an odd prime number	M1	Try dividing 2009 by primes such as 3, 5, 7, 11. If the result is a whole number, see if this is a square number
	$a = 7$	A1	
	$b = 41$	A1	

141 LCM and HCF

Q	Answer	Mark	Comment
1a	$2^3 \times 3$	B1	
1b	$2^2 \times 3 \times 5$	B1	
1c	120	B1	
1d	12	B1	
1ei	$2^4 \times 3^2 \times 5^2$	B1	
1eii	$2^2 \times 3 \times 5$	B1	
2a	$p = 2, q = 3$	B2	
2b	$2^3 \times 3^2 \times 5$	B1	
2c	$a = 2, b = 7$	B2	
2d	$2^2 \times 7^2$	B1	
3ai	1	B1	
3aii	51	B1	
3bi	1	B1	
3bii	pq	B1	

142 Powers

Q	Answer	Mark	Comment
1a	4^8	B1	
1b	6^3	B1	
1ci	4	B1	
1cii	3	B1	
1d	49	B1	
1e	100 000 000	B1	
1f	1 000 000	B1	
2a	x^7	B1	
2b	x^4	B1	
2c	$(ab)^n$	B1	
2d	$(a \div b)^n$	B1	
3a	Trying any number in the formula	M1	
	$a = -0.5$	A1	
3b	Testing a value on both sides	B1	
	Yes and at least one more test	B1	

143 Number skills

Q	Answer	Mark	Comment
1ai	He has forgotten the zero when multiplying by the 4	B1	
1aii	1776	B1	
1bi	He has added instead of multiplying	B1	
1bii	Sight of 203 and 1160	M1	
	1363	A1	
2a	Sight of 576 and 2880	M1	
	3456	A1	

2b	Sight of 392 and 1960	M1	
	£23.52	A1	
3a	261	B1	
3b	1508	B1	
3c	3045	B1	

144　Number skills

Q	Answer	Mark	Comment
1a	29; 58; 290; 580	B2	You will get 1 mark for three correct
1b	Complete chunking method	M1	
	52	A1	
1ci	52 remainder 12	B1	
1cii	26	B1	
2a	Sight of 5472 and 9120	M1	
	14 592	A1	
2b	$912 \div 24$	M1	
	38	A1	
3a	12	B1	
3b	48	B1	
3c	104	B1	
3d	416	B1	

145　Number skills

Q	Answer	Mark	Comment
1a	Sight of 136 and 340	M1	
	476	A1	
1b	$(600 - 476) \div 14$	M1	
	9	A1	
2ai	40 or 10	B1	
2aii	0.007 or $\frac{1}{1000}$ or 0.001	B1	
2bi	23.5	B1	
2bii	23.48	B1	
2biii	23.479	B1	
3	1.93, 2.08, 3.24, 7.86	B3	You will get 2 marks for two or three correct and 1 mark for one correct

146　Decimals

Q	Answer	Mark	Comment
1a	£3.84	B1	
	£4.30	B1	
	£1.10	B1	
	£9.24	B1	
1bi	7.8	B1	
1bii	Sight of 4.8 and 1.44	M1	
	6.24	M1	
1ci	£6.24	B1	
1cii	£300	B1	
2ai	17.4	B1	

2aii	Sight of 9.2 and 4.14	M1	
	13.34	A1	
2b	£13.34	B1	
3	$13 500	B1	
	2500 ÷ 1.65	M1	
	£1515.15	A1	

147 More fractions

Q	Answer	Mark	Comment
1a	$\frac{15}{20} + \frac{8}{20}$	M1	
	$1\frac{3}{20}$	A1	
1b	$\frac{11}{3} - \frac{9}{5}$	M1	
	$\frac{55}{15} - \frac{27}{15}$	A1	
	$1\frac{13}{15}$	A1	
1c	$\frac{2}{5} + \frac{1}{4} + \frac{1}{6}$	M1	
	$\frac{24}{60} + \frac{15}{60} + \frac{10}{60}$	M1	
	$\frac{11}{60}$	A1	
2a	$\frac{5}{2} \times \frac{7}{5}$	M1	
	$3\frac{1}{2}$	B1	
2b	$\frac{33}{10} \div \frac{12}{5}$	M1	
	$\frac{33}{10} \times \frac{5}{12}$	M1	
	$1\frac{3}{8}$	A1	
2c	$\frac{1}{2} \times \frac{9}{10} \times \frac{13}{6}$	M1	
	$\frac{39}{40}$	A1	
3	$\left(1 + \frac{3}{5} + \frac{2}{3}\right)x = 136$	M1	You can also try trial and improvement to solve this
	$x = 136 \div 2\frac{4}{15}$	M1	
	60	A1	

148 More number

Q	Answer	Mark	Comment
1ai	−12	B1	
1aii	−3	B1	
1aiii	−20	B1	
1b	Any two negatives where the second is 2 less than the first, e.g. $-5 - -7$; $-9 - -11$	B1	
1ci	9 is a square number but −4 is not; as all square numbers are positive	B1	
1cii	Total (7) ÷ 7	M1	
	1	A1	
2ai	50	B1	
2aii	0.4	B1	

2b	200	B1	
2c	$\frac{50-40}{0.4}$	M1	
	25	A1	
3a	Any that work such as $3 \times 4 \div -6$	B1	
3b	Any that work such as $-3 \times -4 \div -6$	B1	

149 Ratio

Q	Answer	Mark	Comment
1	$360 \div 18$	M1	
	20	A1	
	160	A1	
2a	$4 : 3$	B1	
2b	$1 : 0.4$	B1	
2c	$1000 \div 8$	M1	
	375 ml	A1	
3	$15 \div 3$	M1	
	35	A1	
4	(F), C, F, F, T	B3	You will lose a mark for every wrong answer

150 Speed and proportion

Q	Answer	Mark	Comment
1a	$72 \div 2.25$	M1	
	32	A1	
	km pr hour	B1	
1b	$72 - 9$	M1	
	$63 \div 1.75$	M1	
	36	A1	
2a	$50 \div 275 \times 165$	M1	
	30	A1	
2b	$275 \div 50 \times 26$	M1	
	143 miles	A1	
3	$600 \div 155$ or $800 \div 220$	M1	
	3.87 or 3.63	A1	
	Handy size as $3.87 > 3.63$	Q1	Make sure you use numbers and write a clear conclusion
4	$20 \div 10$ or $20 \div 40$	M1	
	2 hours and 30 mins	A1	
	No as $40 \div 2.5 = 16$	Q1	Show all your working and use numbers in your conclusion

151 Percentages

Q	Answer	Mark	Comment
1	Decimal / Fraction / Percentage: 0.35, $\frac{7}{20}$, 35; 0.8, $\frac{4}{5}$, 80; 0.9, $\frac{9}{10}$, 90	B3	You will get 2 marks for four or five right and 1 mark for two or three right
1bi	$\frac{3}{10}$	B1	
1bii	70%	B1	

1c	$\frac{11}{20}$ (0.55); $\frac{14}{25}$ (0.56); 57% (0.57); 0.6	B2	You will lose a mark for every wrong answer
2a	$6000 \times 0.88 \times 0.9$	B1	
2b	£6.80 + £3.40 + £1.70	M1	
	£11.90	A1	
3	1.05 seen	B1	
	0.97 seen	B1	
	Both same as $1.05 \times 0.97 = 0.97 \times 1.05$	Q1	Make sure you use numbers and write a clear conclusion

152 Percentages

Q	Answer	Mark	Comment
1a	0.15×350	M1	
	£52.50	A1	
1b	£297.50	B1	
2a	1.175×700	M1	
	£822.50	A1	
2b	0.88×250	M1	
	£220	A1	
3	405	B1	
	$405 \div 2250$	M1	
	18%	A1	
4	1.10 or 0.9 seen	B1	You can also start with any value (say £100) and show that this is £110 then £99
	Because $1.10 \times 0.9 = 0.99$ not 1	B1	

155 Perimeter and area

Q	Answer	Mark	Comment
1a	$78 + 59 + 78 + 59$	M1	
	274 cm	A1	
1b	59×78	M1	
	4602 cm²	A1	
2a	$32-36$	B1	
	km²	B1	
2b	$0.5 \times 4 \times (3 + 6)$	M1	
	18 cm²	A1	
3	$4x = x^2$	M1	Equate the area and the volume
	4	A1	

156 Perimeter and area

Q	Answer	Mark	Comment
1	Dividing shape into two rectangles	M1	You can draw a vertical or horizontal line
	$60 + 40$	A1	
	100 cm²	A1	
2a	$0.5 \times 6 \times 9$	M1	You will need to recall the formula for the area of a triangle
	27 cm²	A1	
2b	$0.5 \times 6 \times (10 + 19)$	M1	The formula for the area of a trapezium is given on the formula sheet
	87 cm²	A1	
3	$2 \times (h - 3) + 24 = 39$	M1	
	$2h = 21$	A1	
	10.5 cm	A1	

157 Area

Q	Answer	Mark	Comment
1a	6×4	M1	You will need to learn the formula for the area of a parallelogram
	24 cm^2	A1	
1b	4×11	M1	
	44 cm^2	A1	
2	$0.5 \times 8 \times (22 + 16)$	M1	
	152 cm^2	A1	
3	Any that work such as $a = 4$, $b = 6$ and $h = 12$	B2	You will get 1 mark for $(a + b) \times h = 60$

158 Symmetry

Q	Answer	Mark	Comment
1a		B2	You will lose a mark for every one wrong or missing
1b		B1	
1ci	0	B1	
1cii	2	B1	
2a		B1	
2b	5	B1	
3a		B1	
3b		B1	

159 Angles

Q	Answer	Mark	Comment
1ai	62°	B1	
1aii	220°	B1	
1b		B1	
2a	$180 - (95 + 48)$	M1	
	37°	A1	
2b	$360 - (98 + 105)$	M1	
	157°	A1	
3	Any three angles, one actute, one obtuse and one reflex that add up to 360. For example: 60, 100, 200	B3	You will lose a mark for any condition that is not met, e.g. no total of 360 or no reflex angle

160   Angles

Q	Answer	Mark	Comment
1ai	Isosceles	B1	
1aii	$p = 72°$, $q = 36°$	B2	
1b	65 + 68	M1	
	133	A1	
2	$(360 − (98 + 36)) ÷ 2$	M1	
	113	A1	
	67°	A1	
3	Trying for example $C = 10$ and finding that this leads to 125°	M1	
	$C = 20$	A1	
	120	A1	

161   Polygons

Q	Answer	Mark	Comment
1a	$p = 45°$, $q = 135°$, $r = 45°$	B3	
1b	Pentagon split into three triangles	M1	
	$3 × 180 = 540$	A1	
2a	$360 ÷ 36$	A1	
	10	A1	
2b	$180 − 160 = 20$	M1	
	$360 ÷ 20 = 18$	A1	
3	$360 ÷ 6$ or $360 ÷ 5$	M1	
	60 and 72	A1	
	132°	A1	

162   Parallel lines and angles

Q	Answer	Mark	Comment
1a	45°	B1	
	Allied or interior angle	B1	
1b	65°	B1	
	Corresponding	B1	
1c	70°	B1	
2	$p = 38$, opposite	B2	
	$q = 38$, alternate	B2	
	$r = 38$, corresponding	B2	
	$p = 142$, interior or allied	B2	
3	Yes as APX is 138 and this is corresponding to CQP	Q2	There are many other answers but you must make your explanation clear

163   Quadrilaterals

Q	Answer	Mark	Comment
1a	Parallelogram	B1	
1bi	Square	B1	
1bii	Rhombus	B1	
2	BCD = 100 (angles in a kite)	B1	If the question asks for reasons, give them or you will lose marks
	DCF = 120 (angles in a parallelogram)	B1	
	DCE = 80 (angles on a straight line)	B1	
	CED = 40 (alternate to ECF)	B1	

3a	All angles 90°	**B1**	
3b	Diagonal cross at right angles	**B1**	
3c	Rotational symmetry of order 2	**B1**	
3d	Any valid answer for pair chosen e.g. Rhombus and square have sides same length	**B1**	When you are given a choice go for something straightforward

164 Bearings

Q	Answer	Mark	Comment
1a	Allow ±0.1 km and ±1° 3.6 km at 034°	**B2**	1 mark for the distance and 1 for the bearing
1b	5.1 km at 169°	**B2**	
1c	5 km at 233°	**B2**	
1d	3.6 km at 304°	**B2**	
2	000° or 360°	**B1**	You will lose a mark for every wrong answer
	090°	**B1**	
	180°	**B1**	
	270°	**B1**	
3a	030° and 340°	**B1**	
3b	210° and 160°	**B1**	
3c	Difference is 180°	**B1**	

165 Circles

Q	Answer	Mark	Comment
1	Clockwise from bottom-left: Tangent	**B1**	
	Radius	**B1**	
	Diameter	**B1**	
	Chord	**B1**	
2a	$\pi \times 12 \times 12$	**M1**	You will need to learn the formula for the area of a circle
	452.4 cm²	**A1**	
2b	$\pi \times 20$	**M1**	You will need to learn the formula for the circumference of a circle. Don't get the area and circumference formulae mixed up
	62.8	**A1**	
3	$0.5 \times 10 \times 10 \times \pi$	**M1**	
	50π	**A1**	
4	$\pi r^2 = 0.5 \times \pi \times 4 \times 4$	**M1**	
	$r^2 = 8$	**A1**	
	2.8 cm	**A1**	

166 Scales

Q	Answer	Mark	Comment
1a	28.4 °C	**B1**	
1b	57	**B1**	
	mph	**B1**	
1c		**B1**	
2	Man ≈ 1.8 m	**B1**	Doorways are about 2 m high
	10–12 m	**B1**	You will get a mark for 6 × whatever height you get for the man

Workbook answers

19

3	Marking values on scales	B1	Note that on both scales every division is 2.5
	135	B1	

167 Scales and drawing

Q	Answer	Mark	Comment
1a	6.7 cm	B1	
1b&c		B2	
1d	5 cm;	B1	
	25 km	B1	
2a	Triangular prism	B1	
2b	1.3 cm	B1	
2c	$1.5 \times 4 = 6$ cm^2	B1	
	$0.5 \times 1.5 \times 1.3 = 0.975$	B1	
	19.5–20 cm^2	B1	
3	40×60	M1	
	2400 m^2	A1	

168 3-D drawing

Q	Answer	Mark	Comment
1a	80 cm^3	B1	
1b		B1	
2		B2	You will get full marks for any isometric view and 1 mark for an isometric view of any five-cube shape
3		B2	Any view will get full marks and you will get 1 mark for any view of a four-cube shape

169 Congruency and tessellations

Q	Answer	Mark	Comment
1a	C; D and F	B1	
1b	B	B1	
2a	A pattern of shapes with no gaps and no overlap	B1	
2b		B1	

3		The angle formed when two octagons are joined is 90° and an octagon does not have a 90° angle to fit into this	B2	You get 1 mark for the diagram and 1 mark for mentioning the angles

170   Transformations

Q	Answer	Mark	Comment
1a	Translation	B1	
	$\begin{pmatrix} -6 \\ -3 \end{pmatrix}$	B1	
1b	Triangle at (3, 6), (3, 9), (5, 6)	B1	
1c	$\begin{pmatrix} -7 \\ 5 \end{pmatrix}$	B1	
2a 2b 2c		B3	
3	Translation $\begin{pmatrix} 3 \\ -3 \end{pmatrix}$	B1	
	Reflection in $y = x$	B1	

171   Transformations

Q	Answer	Mark	Comment
1a 1b		B2	
1c	Clockwise	B1	
	90°	B1	
	Centre (4, 3)	B1	
2a	Enlargement sf $\frac{1}{2}$	B1	
	Centre (1, 8)	B1	
2b	Rays from (1, 0) through corners	M1	
	Triangle at (2, 1), (3, 1), (2, $2\frac{1}{2}$)	B1	

3	 Triangle at (4, 3), (7, 3) and (4, 9)	B1	
	Triangle at (7, 0), (7, 12), (13, 0)	B1	
	Enlargement sf 3, centre (1, 3)	B1	

172 Constructions

Q	Answer	Mark	Comment
1a	Sides 5 cm and 7 cm; included angle 55° 	B3	You will get 1 mark for each side drawn accurately and i mark for the angle
1b		B3	You will get 1 mark for each side drawn accurately and 1 mark for the angle
2a		B2	You will lose a mark if arcs are not shown
2b		B2	You will lose a mark if arcs are not shown
3		B3	You will get 1 mark for a sketch of any triangle with sides 7 and 8 and an angle of 40

173   Constructions and loci

Q	Answer	Mark	Comment
1		B2	You will lose a mark if arcs are not shown
2a	Circle radius 4 cm around flower bed	B1	
2b	Rectangle drawn 1 cm from edge	B1	
	Overlapping area shaded	B1	
3		B4	You will get two marks for the perpendicular bisector and 2 marks for the angle bisector. You will lose a mark if the area is not claerly marked

174   Units

Q	Answer	Mark	Comment
1a	900 cl	B1	
1b	4.5 seen	B1	
	25	B1	
1c	9×20	B1	
	180 g	A1	
2a	$12 \times 30 \div 2.2$	M1	
	164	A1	
	No as $164 > 150$	Q1	Make sure you use numbers in your conclusion
2bi	160 000	B1	
2bii	100	B1	
2c	$100 \div 30 \approx 3.3$	B1	
	4.5 seen	B1	
	Yes as $20 \div 4.5 \approx 4.4 > 3.3$	Q1	Make sure your answer is clear and uses numbers
3	2.2 seen	B1	
	$10 \div 2.2 \approx 4.5$ kg	B1	
	2 h 30 m	B1	

175   Surface area and volume of 3-D shapes

Q	Answer	Mark	Comment
1a	0.2 m	B1	
1b	0.2×0.3	M1	
	0.06 m^2	A1	

2a	$2 \times 4 \times 6$ cm	B1	
2b	12, 8 and 24 seen	M1	
	88 cm^2	A1	
2c	$2 \times 4 \times 6$	M1	
	48 cm^3	A1	
3	Two of 5, 6 and 3 seen	B1	
	$5 \times 6 \times 3$	M1	
	90 cm^3	A1	

176 Prisms

Q	Answer	Mark	Comment
1a	$0.5 \times 4 \times 7$	M1	
	14 cm^2	A1	
1b	14×12	M1	
	168 cm^3	A1	
2	$\pi \times 4 \times 4 \times 10$	M1	
	160π	A1	
3	$117 \div 18$	M1	
	6.5 cm	A1	
4	$\pi \times r^2 \times 2r = 128\pi$	M1	
	$2r^3 = 128$	A1	
	$r = 4$ cm	A1	

177 Pythagoras' theorem

Q	Answer	Mark	Comment
1	$10^2 + 8^2$	M1	
	$\sqrt{164}$	M1	
	12.8	A1	
2	$15^2 - 11^2$	M1	
	$\sqrt{104}$	M1	
	10.2	A1	
3	$2.1^2 + 4^2$	M1	
	$\sqrt{20.41}$	M1	
	4.5	A1	
4	$13^2 - 11^2$	M1	
	$48 + 15^2$	M1	
	$\sqrt{273}$	A1	
	16.5 cm	A1	

180 Basic algebra

Q	Answer	Mark	Comment
1a	$x - 2$	B1	
1b	$2x$	B1	
1c	$2x - 4$ or $2(x - 2)$	B1	
1d	$x + x - 2 + 2x + 2x - 4$	M1	
	$6x - 6$	A1	
2a	$3y \times y = 3y^2,\ 2y + y = 3y,\ 3(y + 1) = 3y + 3$	B3	You will get 1 mark for each answer
2bi	$3q$	B1	

2bii	$15pq$	B1	
2biii	$9x - 4$	B1	
3a	Values that work, e.g. $a = 3$, $b = 2$, $21 - 6 = 15$	B1	
3b	Values that work, e.g. $a = 2$, $b = 1$, $14 - 3 = 11$	B1	
3c	Values that work, e.g. $a = 2$, $b = 4$, $14 - 12 = 2$	B1	Both values need to be even or both need to be odd

181　Expanding and factorising

Q	Answer	Mark	Comment
1a	$5x - 15$	B1	
1b	$2x + 2 + 6x + 4$	M1	
	$8x + 6$	A1	
1c	$2(2x + 3) + 2(x + 3)$	M1	
	$6x + 12$	A1	
2a	$3x - 12 + 8x + 2$	M1	
	$11x + 10$	A1	
2bi	$2(2x + 3)$	B1	
2bii	$x(5x + 2)$	B1	
3ai	$6x^2 - 8xy - x^2 - 3xy$	M1	
	$5x^2 - 11xy$	A1	
3b	$12x - 18y - 2x + 6y$	M1	
	$10x - 12y$	A1	
4a	$4(3x + 1) - 2(2x - 1)$	M1	
4b	$8x + 6$	A1	

182　Linear equations

Q	Answer	Mark	Comment
1a	$\frac{x}{3} = 5 - 4$	M1	
	3	A1	
1b	$3x = 1 - 4$	M1	
	-1	A1	
1c	$7x - 2 = 12$	M1	
	$7x = 14$	A1	
	2	A1	
2a	$2x + 5 = x + 6$	M1	
	$2x - x = 6 - 5$	A1	
	$x = 1$	A1	
2b	$3y - 6 = y + 5$	M1	
	$3y - y = 5 + 6$	A1	
	$y = 5.5$	A1	
3a	$2x = 16$	M1	
	$x = 8$	A1	
3b	$2(x + 5) = 8$ expands to $2x + 10 = 8$ and the solution is $x = -1$	Q2	Make sure you explain what the brackets expand to and solve the equation
4	$2y + 3 = \frac{y}{2} - 3$	M1	
	$y = -4$	A1	
	$x = -5$	A1	

183 Linear equations

Q	Answer	Mark	Comment
1ai	$3x + 5 = 26$	B1	
1aii	$3x = 21$	M1	
	$x = 7$	A1	
1bi	$12y - 8 = 16$	M1	
	$12y = 24$	M1	
	$y = 2$	A1	
1bii	$5x - x = 10 + 2$	M1	
	$4x = 12$	A1	
	$x = 3$	A1	
2a	$5x - 3x = 1 + 2$	M1	
	$2x = 3$	A1	
	1.5	A1	
2b	$3x + 12 = x - 5$	M1	
	$2x = -17$	A1	
	-8.5	A1	
2c	$5x - 10 = 2x + 8$	M1	
	$3x = 18$	A1	
	6	A1	
3a	$4x + 12 = x + 3$	M1	
	$4x - x = 3 - 12$	M1	
	$x = -3$	A1	
3b	$10x - 5 = 2x - 6$	M1	
	$10x - 2x = -6 + 5$	M1	
	$x = -\frac{1}{8}$	A1	
4	Sight of $5a + 2b = 16$	M1	
	$a = 2$	A1	
	$b = 3$	A1	

184 Linear equations

Q	Answer	Mark	Comment
1a	$a = 4$	B1	
	$b = 6$	B1	
	$c = 7$	B1	
	$d = 8$	B1	
1bi	$50x + 140$	B1	
1bii	$50x + 140 = 340$	B1	
	$x = 4$	A1	
2	$x + 3x - 1 + 2x + 5 = 25$	M1	
	$6x + 4 = 25$	A1	
	3.5 cm	A1	
3	$8x + 8$ or $11x + 9$	B1	
	$10x = 9x + 17$	M1	
	17	A1	

185   Trial and improvement

Q	Answer	Mark	Comment
1	(5.6 =) 198.016	M1	
	(5.7 =) 207.993	M1	
	Testing 5.65 (202.96) and stating 5.7 as the answer	A1	
2	Testing 4 (= 8.5)	M1	
	(3.7 =) 7.94	M1	
	(3.8 =) 8.12	M1	
	Testing 3.75 (8.033) and stating 3.7 as the answer	A1	
3	$x^2(x + 6) = 2000$	M1	
	(10.8 =) 1959.552	M1	
	(10.9 =) 2007.889	A1	
	Testing 10.85 (1983) and stating 10.9 as the answer	A1	

186   Formulae

Q	Answer	Mark	Comment
1a	$x + 6$	B1	
1b	$4x + 6$	B1	
1c	$4x + 6 = 27$	M1	
	5.25 g	A1	
2a	4	B1	
2b	52	B1	
2ci	$200f + 50e$	B1	
2cii	£10 000	B1	
3a	$x = \frac{C}{\pi}$	B1	
3b	$3x = 6y + 9$	M1	
	$x = 2y + 3$	A1	
4a	$10y = 6x + 124$	M1	
	$y = \frac{6x + 124}{10}$	A1	
4b	$y = \frac{6 \times 12 - 124}{10}$	M1	
	25p	A1	

187   Inequalities

Q	Answer	Mark	Comment
1a	$3x \leqslant 6$	M1	
	$x \leqslant 2$	A1	
1b	$x > -2$	B1	
1c	$-1, 0, 1, 2$	B1	
2a	$-3 < x \leqslant 1$	B1	
2bi	$\frac{x}{2} > -2$	M1	
	$x > -4$	A1	
2bii	$x + 3 \leqslant 2$	M1	
	$x \leqslant -1$	A1	
2c	$-3, -2, -1$	B1	
3	3, 5, 7, 9	B2	You will lose a mark for every missing value

188   Graphs

Q	Answer	Mark	Comment
1a	3 miles	B1	
1b	13 km	B1	
1c	100 miles	B1	
2ai	2 km	B1	
2aii	5 minutes	B1	
2bi	40 minutes	B1	
2bii	5 km in 40 mins	M1	
	7.5 km/h	A1	
3	Line from (10, 0) to (12.30, 30)	B1	
	Line from (11, 0) to (11.30, 10) to (11.45, 10) to (12.45, 30)	B1	
	Jogger by 15 min	B1	

189   Linear graphs

Q	Answer	Mark	Comment
1a	1; 4; 7	B1	
1b		B2	You get 1 mark for a line passing through (0, 1) and 1 mark for a line with a gradient of 3
1c	2.3	B1	
2		B2	You will get a mark for a line passing through (0, −1) and a mark for a line from (−3, −7) to (3, 5)
3a	Line from (0, 20) to (4, 140)	B1	
	Line from (0, 30) to (4, 130)	B1	
3b	Same at 2 hours	B1	
	Bernice as she is cheaper from 2 to 4 hours	B1	

190	Gradient-intercept		
Q	Answer	Mark	Comment
1a	D	B1	
1b	C	B1	
1c	D and E	B1	
1d	$3; \frac{1}{2}; -\frac{5}{3}$	B3	You get 1 mark for each correct answer
2	Graph intersecting y-axis at -2	B1	
	Graph with gradient of 3	B1	
3ai	$x + y = -4$	B1	
3aii	$x = -3$	B1	
3b	-7	B1	

191	Quadratic graphs		
Q	Answer	Mark	Comment
1a	4, 3, 4, 7	B1	
1b	Graph of $y = x^2 + 3$	B2	You will get 1 mark for plotting the points and 1 mark for a smooth curve
2a	$-4, -6, -4$	B1	
2b	Graph of $y = x^2 - 3x - 4$	B2	
3	A is $y = x^2 + 3$; B is $y = x^2$; C is $y = x^2 - 4$	B2	You will get 1 mark for one correct

192 Quadratic graphs

Q	Answer	Mark	Comment
1a	0, 1, 4, 9	B1	
1b	Graph of $y = x^2 - 2x + 1$	B2	You will get 1 mark for plotting the points and 1 mark for a smooth curve
1c	−1.45, 3.45	B1	
1d	1	B1	
2a	7, −2, −1, 2	B1	
2b	Graph of $y = x^2 + 2x - 1$	B2	
2c	−2.9, 0.9	B1	
2d	−2.4, 0.4	B1	
3	P(0, 0), Q(4, 0), R(1, −3) S(3, −3), T(2, −4)	B3	You will lose a mark for every wrong answer

193 Pattern

Q	Answer	Mark	Comment
1a	$1 + 3 + 5 + 7 = 16 = 4^2$	B1	
	$1 + 3 + 5 + 7 + 9 = 25 = 5^2$	B1	
1b	99	B1	
1c	225	B1	
2a		B1	
2b	6, 10, 15	B2	You lose a mark for every wrong answer
2c	The number added goes up by 1 more each time	B1	
3	401 seen	B1	
	200×401	M1	
	80 200	A1	

194   The *n*th term

Q	Answer	Mark	Comment
1a	5, 9, 13	B1	
1b	7th	B1	
1c	The terms are all odd numbers, 84 is not odd	B1	
1d	$7n - 4$	B2	You will get 1 mark for writing down $7n$
2a	16, 21, 26	B2	
2b	101	B1	
2c	$5n + 1$	B2	You will get 1 mark for writing down $5n$
3a	5, 7, 9, 11, 13, 15, 17, 19, 21, 23 and 4, 9, 14, 19, 24, 29, 34, 39, 44, 49	B2	You get 1 mark for each list
3b	9, 19, 29, 39, …	M1	
	$10n - 1$	A1	

195   Sequences

Q	Answer	Mark	Comment
1a	Always odd	B1	
1b	Always even	B1	
1c	Could be either	B1	
1d	Always odd	B1	
1e	Always odd	B1	
2a	$2 \times$ anything is even	B1	
2b	$2n \times 2n$	M1	
	$4n^2$ which is a multiple of 4	A1	
3a	18, 24, 30, 36	B2	You will lose a mark for each error
3b	$6n + 6$ or $6(n + 1)$	B2	You wil get 1 mark for $6n$
4	$96 - 5n$	B1	
	$96 - 5n = 4n - 3$	M1	
	11th term, 41	A1	

Notes

Student notes